T0295583

R&D INVESTMENT OF MULTINATIONAL CORPORATIONS AND CHINA'S INDEPENDENT INNOVATION

Series on Innovation and Operations Management for Chinese Enterprises

Print ISSN: 2591-7188
Online ISSN: 2591-7196

Editor: Xiaobo Wu *(School of Management, Zhejiang University, China)*

Published:

Series on Innovation and Operations Management
for Chinese Enterprises – Vol. 5

R&D INVESTMENT OF MULTINATIONAL CORPORATIONS AND CHINA'S INDEPENDENT INNOVATION

Wen Xiao
Jiadong Pan
Gaobang Lin
Zhejiang University, China

 World Scientific

 ZHEJIANG UNIVERSITY PRESS
浙江大学出版社

Published by

World Scientific Publishing Co. Pte. Ltd.
5 Toh Tuck Link, Singapore 596224
USA office: 27 Warren Street, Suite 401-402, Hackensack, NJ 07601
UK office: 57 Shelton Street, Covent Garden, London WC2H 9HE

and

Zhejiang University Press
No. 148, Tianmushan Road
Xixi Campus of Zhejiang University
Hangzhou 310028, China

Library of Congress Cataloging-in-Publication Data
Names: Xiao, Wen (Economist), author. | Pan, Jiadong, author. | Lin, Gaobang, author.
Title: R&D investment of multinational corporations and China's independent innovation /
    Wen Xiao, Jiadong Pan, Gaobang Lin, Zhejiang University, China.
Other titles: R & D investment of multinational corporations and China's independent innovation
Description: [Singapore] : World Scientific, [2020] | Series: Series on innovation and operations
    management for Chinese enterprises, 2591-7188 ; vol. 5 |
    Includes bibliographical references and index.
Identifiers: LCCN 2020032972 | ISBN 9789811220890 (hardcover) |
    ISBN 9789811221491 (ebook) | ISBN 9789811221507 (ebook other)
Subjects: LCSH: Research, Industrial--China. | Endowment of research--China. | Globalization.
Classification: LCC T177.C5 X56 2020 | DDC 607.2/51--dc23
LC record available at https://lccn.loc.gov/2020032972

British Library Cataloguing-in-Publication Data
A catalogue record for this book is available from the British Library.

For any available supplementary material, please visit
https://www.worldscientific.com/worldscibooks/10.1142/11846#t=suppl

Desk Editors: George Vasu/Lum Pui Yee

Typeset by Stallion Press
Email: enquiries@stallionpress.com

Preface

Although research and development (R&D) internationalization is a concept defined almost a decade ago, R&D internationalization of multinational corporations (MNCs) is a recent phenomenon that is drawing increasing attention from the academic community. MNCs enjoy advantages in monopoly, location and internalization as they make full use of the imperfection of markets of all the countries through international division of labor within the corporations and "internalize" those markets with a strict hierarchical organization. In terms of R&D, technology, which embodies the stock of knowledge and production know-how of an organization, cannot be packaged, sold, or even recorded in most cases. Organizations prefer to transfer their core competitiveness — technology — internally through direct investment in pursuit of maximum profit. This process avoids and eliminates the risks involved in trading with outsiders, reduces intervention from host governments and transcends the trade barriers. In this way, technology seems inaccessible to the outsiders, which prevents others from research in a comprehensive and systematic manner. In this book, technology is regarded as a factor of production that is allocated by MNCs across the world to maximize profit. Such a rational economic behavior serves as our entry point of research, and technology trade diagrams are used in this book to enrich the existing theoretical results of R&D internationalization.

Generally speaking, there are three ways to achieve progress in technology: indigenous innovation, technology transfer and R&D internationalization. Among them, R&D internationalization did not receive much attention until the recent years. China has attracted a huge amount of investment in R&D internationalization of MNCs, which is playing an important role in its economy. Constrained by the natural resources, China has to attach more importance to technological progress and indigenous innovation in order to go beyond the old economic growth model. Currently, the increase in R&D investment from MNCs across China is mainly attributed to the huge share of the Chinese market and the relatively cheap manpower. Meanwhile, an export-oriented economy is the major growth booster for China. As a typical representative in processing trade, China has built its fast-growing economy upon high energy consumption. This kind of extensive growth is driven more by increase in input than by technological progress or indigenous innovation, as evidenced by the one-time level effect rather than the sustainable effect. In the long term, only by continuously improving indigenous innovation capacity and boosting technological progress can there be sustainable economic growth. In this sense, the R&D investment of MNCs can help increase China's indigenous innovation capacity and promote its economic transformation and upgrading.

However, while MNCs are increasing their R&D investment, China is actually recording decrease in overall competitiveness. This has given rise to six questions: How do MNCs make decisions for R&D internationalization? Does the investment in R&D have technological spillover effects on China's indigenous innovation capacity? What are the channels for such effects to affect the domestic enterprises in China? How can China make better use of the spillover effects brought about by the R&D investment of MNCs when it is restrained by limited natural resources? What is the relation between China's economic transformation and the regional indigenous innovation capacity? How do R&D internationalization of MNCs and the indigenous innovation capacity of China facilitate China's economic transformation and upgrading?

In this book, we answer the above questions through systematic analysis and demonstration, and shed light on how R&D internationalization enhances China's indigenous innovation capacity and boosts China's economic transformation and upgrading. First of all, we review the classic theories on R&D internationalization of MNCs and compare the existing theories from four perspectives: driver analysis, selection of location, the effects of outward transfer of technology and the system of technology transfer. We find that R&D internationalization of MNCs is not designed to improve the indigenous innovation capacity of host countries; instead, the improvement in such capacity is the result of the objective technological spillovers during MNCs' pursuit of maximum profit around the globe. We also build an innovative indicator system to evaluate the overall capacity of regional indigenous innovation, measure such capacity in China, delve into the evolutionary characteristics of the overall capacity of indigenous innovation of regions in China in the recent years and find out the major locational features.

On this basis, we figure out the regional level of R&D internationalization in China and its efficiency by using principal component analysis and data envelopment analysis and examine the intrinsic connection between the two. Meanwhile, we analyze five major influencing factors: regional market size, scale of foreign investment, level of protection of intellectual property rights, internal R&D resources and infrastructure construction. It is found that when there is a larger market, more foreign investment, better protection of intellectual property rights and richer R&D resources, it is easier for pure technical efficiency and scale efficiency to promote R&D internationalization; otherwise, the contribution to technological progress is greater.

Then, we explore how R&D internationalization facilitates indigenous innovation at the theoretical level, in terms of the technology spillover effects of R&D internationalization, the technology transfer system, the progression mechanism of indigenous innovation and imitative innovation. We build a classic growth model to derive the long-term and stable equilibrium solution to the growth of MNCs and host countries. The improvement in the indigenous innovation capacity of host countries is approached from three angles: the interaction between "introduction" and "innovation", the progress from "imitation" to "innovation" and the shift from "imitation" to "innovation". Then, we probe into the inhibitory effect of R&D internationalization on indigenous innovation with focus on the inhibitory effect of the one-way flow of professionals, the crowding-out effect of competition and the defensive follow-up strategy. Meanwhile, we adopt the paradigm of institutional economics to analyze the inhibitory effect and point out that the institutional inhibitory effect of R&D internationalization can be avoided through institutional innovation.

We move on to investigate how R&D internationalization enhances China's indigenous innovation capacity based on quantitative analysis and case studies. It is found that increasing R&D internationalization can actively promote the improvement in regional indigenous innovation capacity, yet with a threshold effect in terms of regional distribution.

To be specific, such a phenomenon is more apparent in the provinces in Eastern China than those in Central and Western China. Moreover, quantitative analysis reveals that, in regions with a larger market, more foreign investment, better protection of intellectual property rights and better infrastructure, R&D internationalization plays a stronger role in promoting indigenous innovation capacity.

We also derive and extend Coe and Helpman's model of R&D spillovers, based on which empirical tests should be carried out. As the result indicates, the spillovers from accumulation of R&D capital and import channels of capital goods exert a more apparent and stronger positive influence on the technological progress in China, followed by foreign direct investment (FDI) and FDI in R&D in China. Import of consumer goods, China's overseas direct investment and technology introduction contract do not have an evident influence.

Lastly, based on normative and empirical study, we look into the route choices, ideas and measures for China to enhance indigenous innovation capacity. We believe that this goal can be achieved by joining international alliances of technological innovation, breaking out from technology lock-in by virtue of oligopolistic reaction, encouraging enterprises to be agents of technological innovation, and advancing the building of a national system for technological innovation.

The book is structured as follows: Chapter 1 builds the theoretical framework of the whole book; Chapter 2 discusses the development of internationalization of MNCs; Chapter 3 focuses on China's innovation capability; Chapter 4 studies how R&D investment affects China's innovation; Chapter 5 analyzes the crowding-out effect and defensive follow-up strategy of R&D investment; Chapter 6 carries out an empirical analysis of the knowledge spillovers of overseas R&D capital; Chapter 7 introduces the readers to different ways of R&D investment; Chapter 8 showcases some case studies on the air separation industry in China; Chapter 9 concludes with some important policy suggestions for the enhancement of indigenous innovation capacities of China from the macro-policy and micro-reaction perspectives.

About the Authors

Wen Xiao is a Professor from the School of Economy and the Center for Research Regional Coordination Development, Zhejiang University. Her research interests include international investment and Chinese entrepreneurial behavior. She has published widely in SSCI and other leading economic journals, and has also been involved in a lot of research programs, including the Key National Social Science Foundation of China, the Project of National Natural Science Fund, China's Educational Ministry Key Program, and Zhejiang Science & Technology Plan for financial support. She has received several prestigious awards from the Chinese Education Ministry, the Zhejiang Government and the Zhejiang Education Bureau.

Jiadong Pan is a Lecturer from the Zhejiang Institute of Administration. His research interests include international economy and trade, and regional economy. He has published widely in SSCI and other leading economic journals, and has also participated in various national projects. He has received many key awards from the Zhejiang Institute of Administration.

Gaobang Lin is a Research Scholar from the School of Economy, Zhejiang University. His research interests include innovation and international investment. He has published extensively in leading economic journals. He has played an important role in the Project of National Natural Science Fund and has also received several prestigious awards from the Zhejiang University, the Zhejiang Government and the Zhejiang Education Bureau.

Acknowledgments

This book is the fruit of a program supported by the National Philosophy and Social Science Fund and titled "Economic Research on the R&D Internationalization of Multinational Corporations and Enhancement of China's Indigenous Innovation Capacity" (No. 07BJL029). It is also supported by the Phase III Fund for Program 985 from the Center for Research of Private Economy of Zhejiang University. We have devoted many years of hard work to the preparation of this book. In order to reflect the latest development of R&D internationalization of MNCs and indigenous innovation, we have consulted a large number of relevant literatures during manuscript preparation. These literatures have made important contributions to the study in this book and we have included all these works in the references section. We expect this book to be a new shining point among the vast number of publications on this subject and we would like to express our gratitude to the authors of the various works referenced from the literature.

During the preparation of this book, many scholars, entrepreneurs and government staff have offered us support and help. In particular, Professor Shi Jinchuan and Professor Jin Xiangrong from School of Economics, Zhejiang University have given their strong support for the publication of this book. Thanks are due to Professor Huang Xianhai from School of Economics, Zhejiang University, Professor Zhang Xukun from Zhejiang Gongshang University, Professor Chen Zhiqi from the University of Western Ontario, Canada, and Dr. Liu Jixun and Dr. Zhou Minghai, who have given important suggestions on the main theoretical models and empirical models of this book. Ms. Hou Tian has participated in drafting the fourth section of Chapter 4; the entrepreneurs of enterprises, such as Volks, Wanxiang Group, Transfer Group, Giant KONE Group, Dehong Electric and Linde, have cooperated with us and supported us during the research. We would also like to thank the leaders of Zhejiang Provincial Development and Reform Commission, Shanghai Science and Technology Commission, Zhejiang Science and Technology Department, Huzhou Science and Technology Bureau, Zhongguancun Science Park, Shanghai Pudong Economic and Technological Development Zone, who have offered guidance and support for the preparation of this book, due to which our field research went on smoothly. Readers and colleagues in the academic community are welcome to point out to us any shortcomings in this book.

Contents

Chapter 1

Theoretical Research on the R&D Internationalization of MNCs

The accelerating economic globalization today is boosting the international flow of products and factors. Among these factors, capital and technology, in particular, have had a huge impact on the global economy. As one of the major manifestations of global economic integration, multinational corporations (MNCs) are emerging. They have brought along international flows of capital, technology and products, which have greatly boosted the comovement of economic growth. For any economy in the world, long-term growth of total factor productivity (TFP) is the guarantee for sustainable, rapid and stable economic growth. However, what TFP has contributed to the economic growth in China is a far cry from that contributed in many other countries and is a one-off level effect rather than a sustainable effect. It can be said that TFP growth stems from a country's research and development (R&D) capacity and its ability to absorb the spillovers of the international R&D capital. However, China has been lagging behind developed countries in the West in terms of R&D investment. Given the limited overall R&D capacity in China, it is essential to absorb the spillover effects of international R&D for the improvement of domestic indigenous innovation capacity. The international flow of R&D capital is realized through the R&D internationalization of MNCs.

MNCs tend to transfer only the low-value-added part of the value chain abroad and keep the R&D activities, which is their core competitive advantage, at home. By the end of the 1980s, only very few MNCs had been engaged in global research and development. After the 1990s,

1

the global R&D of MNCs began to flourish along with economic globalization. Many European MNCs became pioneers in promoting R&D internationalization, including ABB in Switzerland, Philips in Netherlands and Ericsson in Sweden. Later, famous MNCs in the United States, Japan and Germany, such as General Electric, General Motor, Toyota, Fujitsu and Daimler, also joined the trend. In 2009, the Department for Innovation, Universities and Skills (DIUS) of UK released the *2008 R&D Scoreboard.* The report covers 1,400 top-notch enterprises in the world and reveals that their R&D investments added up to around 520 billion dollars, up by 9.5% compared with the previous year. It should be noted that MNCs engage in overseas R&D activities not simply to adapt to local needs but to target the global market and to integrate with the knowledge-creating process. As China furthers its reform and opening-up and optimizes its business environment, an increasing number of MNCs choose to set up offices in China and make investments, which is conducive to improving the R&D level of China.

1. Review of the Theoretical Research on R&D Internationalization

As pacemakers of technological innovation, MNCs are developing new strategies to transfer their technological assets to other parts of the world where they are combined with local factors, and join new alliances to reorganize production, in addition to heavy investment in technological innovation (Li, 2004). In 1994, the Canada-based company Northern Telecom Limited funded the BUPT-Northern Telecom Center, which officially kicked started the R&D internationalization of MNCs in China. Since then, an increasing number of MNCs have been establishing R&D centers in China. After China's entry to the WTO, there was an exponential growth in MNCs' investment in R&D institutions across China. For many years, there has been an exponential growth in the R&D internationalization of MNCs. What are the driving factors behind it? What is their interaction with the host countries? This chapter reviews the theories on R&D internationalization of MNCs in four categories, namely, investment in R&D internationalization of MNCs, motivations behind R&D internationalization of MNCs, selection of location and technology spillover.

MNCs tend to have special motivations for their overseas R&D investment, which is different from general productive investment.

Many scholars in the West have probed into such motivations and proposed various theories. In the following section, R&D internationalization of MNCs is introduced and reviewed based on the most representative studies to offer an outline of the related research development.

1.1. *Studies on R&D internationalization based on traditional MNC theories*

For a long time, the academic community has attached great importance to FDI, which has yielded a large number of theoretical achievements. Major traditional FDI theories include the monopolistic advantage theory, the internal optimization theory and the eclectic paradigm of international production, which have empowered many scholars to elaborate on R&D internationalization of MNCs.

1.1.1. *Studies on R&D internationalization of MNCs based on monopolistic advantage theory*

The earliest scholars studied foreign investment of MNCs from the perspective of monopolistic advantage. Hymer (1960) was the first to distinguish FDI from other kinds of investment. He used the concept "monopoly" to examine the FDI of MNCs and coined the term "monopolistic advantage". His theory was enriched and extended by later scholars and became the classic "monopolistic advantage theory" in the FDI theories.

According to the theory, unlike domestic enterprises of host countries, MNCs face unfavorable factors when they organize production overseas as they know little about the local consumption habits, organization and legal system. However, they still choose FDI and outperform the domestic enterprises in the fierce market competition. The key to their success in investment is some kind of monopolistic advantage that offsets the intrinsic unfavorable factors of foreign investment.

Hymer believed that technological advantage is the most important monopolistic advantage of MNCs. Equipped with ample resources, MNCs are able to invest a huge amount of capital in R&D and boast of a strong R&D capacity. They prefer to keep advanced technologies within the company rather than transfer them so as to secure their monopoly. They enjoy monopolistic advantages because of their core assets, of which technology and knowledge are the most important components. The key to monopolistic advantages is MNCs' ability to develop diverse business capabilities based on the core assets, i.e., their ability to differentiate products with technologies and knowledge. The monopolistic advantages of MNCs are public goods that can be shared by the subsidiaries without causing any additional cost to either the parent companies or the subsidiaries. In this sense, knowledge assets are the most typical public goods that form the monopolistic advantages of MNCs and are produced in the activities of MNCs, which derive their monopolistic advantages from the advantages brought about by the knowledge they produce.

The marrow of this theory is that MNCs have the products and technologies for imperfect competition that bring monopolistic advantages and lead to imperfect competition. When the imperfect competition theory is introduced into FDI studies, the basic conclusion is that monopoly and advantage together form the basic motivation for MNCs to make FDI. A major source of advantage for MNCs is their R&D activities. Hence, overseas R&D investment is made to strengthen their monopolistic advantages for greater profit in the international market. The technological advantage of their home countries is not undermined as the R&D activities for core technologies are still conducted at home. The monopolistic advantage theory offers part of the reasons why MNCs make overseas R&D investment, but fails to explain the related industrial characteristics and national models. Besides, it is not convincing when it comes to the FDI of enterprises in developing countries and investment of small and medium-sized enterprises in developed countries.

1.1.2. *Studies on R&D internationalization of MNCs based on technology life cycle theory*

American economist Vernon (1966) proposed the technology life cycle and technology transfer model through empirical study on overseas direct investment of American enterprises. According to the model, technology shares with all organisms a life cycle of generation, growth, decline and death. This process is called the technology life cycle, which can be divided into two stages, namely, technology R&D and technology diffusion and transfer, depending on whether the technology has entered the market. In the first stage, the technology R&D stage, the life of the technology starts with innovation and conception, which echoes the second stage of the creation theory — gestation. The process from conception through invention (including products and methods) to the industrial production, trial sale and launch of the innovative technologies and their products is usually called the R&D process. In the second stage, the stage of technology diffusion and transfer, after the innovative technologies and their products enter the market, they experience investment, growth, maturity, decline, obsolescence and elimination. This process is the market life of technology. The length of this stage is mainly decided by how advanced the technologies are, their growth and economy, and the demand of the international market. Pearce (1989) used the original technology life cycle theory and his revised theory to distinguish the role of overseas research institutions of MNCs. Based on the original theory, he believed that the overseas R&D of MNCs is based on their overall strategy and, in the second stage of the life cycle, is carried out with the outward shift of the productive institutions. In addition, the major function of the overseas R&D institutions is transferring technology to assist subsidiaries with their operations. Pearce argued that the major potential function of MNCs' overseas R&D is to coordinate research institutions in different countries and acquire the innovative technologies of host countries.

Ronstadt (1977) examined the then overseas R&D activities of MNCs and discovered that the mission of such activities was to receive the technologies transferred from their parent companies and the applications of such technologies. This is in line with the product life cycle theory, which holds that MNCs produce and develop technologies in home countries and transfer the technologies internally to their subsidiaries for localization and adaptation to the local market.

The technology life cycle theory is based on the monopolistic advantage theory. It focuses less on the technical support from overseas R&D institutions for foreign productive enterprises but more on the importance of acquiring technological resources worldwide for the overall interests of the company. This theory explains the overseas investment and innovation of the American MNCs for the period between WWII and the 1970s. However, it fails to explain the FDI in R&D of MNCs.

1.1.3. *Studies on R&D internationalization of MNCs based on internalization theory*

In their 1976 work, *The Future of Multinational Enterprise*, British scholars Buckley and Casson put forward the internalization theory, which holds that the transaction failure of the market leads to an increase in transaction costs and prompts MNCs to engage in internalization of transaction. The reasons behind market internalization include difficulty in transaction of intermediate products, failure of external market and high cost in external market transaction. When internalization goes beyond borders, MNCs come into being. In this sense, international market internalization is the process of making FDI. In other words, market internalization drives enterprises to make FDI. Similarly, the R&D activities of MNCs have a close connection with the headquarters, and the productive institutions are also the result of internalization. American scholar Magee (1977) asserted that product development, technology renewal and market dynamics all fall under the category of information, which enjoys the attribute of public goods and can be shared. The patent system plays a limited role in protecting technology. Hence, to ensure more benefits from their investment in information, enterprises have to internalize existing information through FDI to obtain sufficient reward. By setting up R&D departments in host countries, MNCs can directly access the information about technology and market to make prompt responses and adjustments, and have face-to-face interactions with the departments of management, production and marketing for communication of information. This theory to some extent explicates the rapid R&D internationalization of MNCs after the war.

1.1.4. *Studies on R&D internationalization of MNCs based on eclectic paradigm of international production*

British economist John Dunning (1977) combined the strengths of the existing theories and proposed the eclectic paradigm of international production, which has had the greatest impact on the behaviors of MNCs among the theories. Dunning argued that enterprises must be equipped with three advantages, namely, ownership advantages, internalization advantages and location advantages, in order to engage in FDI. Ownership advantages are divided into technology advantage, advantage in business scale and managerial advantage, and advantage in financing capacity. Among them, technology advantage is the most important component, which in the broad sense consists of production techniques, patent, exclusive technologies, organizational and managerial skills, sales techniques, R&D capacity as well as a major direction — the capability of product differentiation after standardized production. Dunning believed that ownership advantages alone cannot effectively explain the direct investment of an enterprise that can resort to other means to extend such advantages. Internalization is a means for MNCs to use their assets internally so as to avoid the negative impact from the imperfection of the external market, realize the optimal allocation of resources, and maintain and make full use of the monopoly of the ownership advantages. The internalization theory highlights the imperfection of intermediate products like knowledge. Intermediate products refer to tangible products like semi-finished raw materials and components as well as knowledge products like technology, patent, managerial skills and market information. Given the special nature of knowledge products, their market structure and their important position in modern corporate management, MNCs' motivation to internalize the knowledge product market is strongest. Since the research and development of technology cost a lot of time and money and knowledge products are "naturally monopolistic" for a certain period of time, they should be used through price discrimination. The uniqueness of knowledge products and their benefits can only be determined after they are put into production, which leaves buyers in uncertainty. Knowledge products can be shared and only their right of use is transferred in transaction, which prevents both sides from guaranteeing that the products will not be sold to a third party. Market transaction of knowledge products also involves the risk of disclosure, which is also a kind of transaction cost. For those reasons, internal transaction of knowledge products is imperative for MNCs. In other words, they have to restrict the use of knowledge products within the company.

Dunning also pointed out that, when MNCs believe that they have certain ownership advantages over their competitors and it is possible to obtain ideal benefits through internalized R&D in overseas locations, they will engage in FDI in R&D. He asserted that the FDI in R&D helps MNCs gain some kind of advantage which, when integrated with their existing advantages, helps them maintain and strengthen their competitiveness. The FDI in R&D of MNCs has one or more of the following functions: (a) adapting or improving products, raw materials or technologies; (b) studying basic materials or products; (c) rationalizing production and research or minimizing their cost; and (d) learning about and monitoring the development of and changes in the technology of other countries. The R&D institutions set up by MNCs from developed countries in the developing world for technological localization are market oriented. They develop new technologies and products based on the market of the host countries in order to grab it. The R&D investment of MNCs in the emerging industrial countries is resource oriented as it is to a large extent designed to make use of their cheap R&D manpower. The mutual investment between developed countries is efficiency oriented because they make use of each other's R&D facilities and technological foundations and benefit from technical specialization to become international technological innovation centers. With investment of MNCs of developed countries as the starting point, this theory cannot explain the external investment from developing countries.

1.2. Basic theoretical research on R&D internationalization of MNCs

Since the mid-1980s, economic globalization has accelerated, FDI has become progressively active and technology is playing an increasingly prominent core role in economic activities. In this book, we call the theories on R&D internationalization of MNCs, in a new context, "contemporary theories", which are divided into the following categories.

1.2.1. Strategic R&D investment theory

Veugelers (1995) put forward the strategic advantage theory based on game theory and behavioral economics. It becomes strategic R&D investment theory when applied to the study of R&D internationalization of MNCs. The theory holds that in the era of globalization, MNCs make foreign investment not simply to maximize profit for the current period but to take up as large a market share as possible to secure their long-term strategic advantage. R&D internationalization is the ultimate choice of their internal integration posture and external coordination for their global operations. They make strategic investment decisions more from the perspective of global strategic competition, with full consideration for the strategic reaction of their competitors, than from the perspective of profitability.

R&D is the most important strategic operation of MNCs and the most important part of the corporate value chain. R&D globalization in effect mirrors two major strategic variables for MNCs in international competition. One variable is the internal integration posture of global operations of MNCs. In other words, MNCs deploy their R&D activities in different areas to achieve optimal integrated advantages. The other variable is external coordination posture, i.e., MNCs' strategic response to the external environment and the level of coordination of the operations. R&D globalization, the result of the interplay of the two variables, allows MNCs to strengthen their strategic advantage and enhance their international competitive status.

Strategic R&D investment theory explains why the mutual competitive investment between developed countries is heightening. Especially after the 1990s, MNCs have been fighting fiercely for the emerging markets.

1.2.2. *Need–Resource relation theory*

The need–resource theory (NR theory) was proposed by Saito Yu in 1979. To be specific, during the R&D process, there is first of all some kind of need for developing a new technology (N: NEED). Then, necessary resources are invested (R: RESOURCE), including professionals, funds, equipment, organization and information. The subjects of technology development then make creative use of R to meet N in the development of technologies. In other words, R&D is realized with creativity as the critical juncture in the interplay between N and R.

For MNCs, the relation between countries in terms of needs and resources forms the motivation behind the foreign R&D investment. Saito Yu believed that the internal foundation for R&D internationalization is the internationalization of the N–R relation, which is the core of the R&D model. To adapt to global competition, it is necessary for MNCs to allocate existing R&D resources worldwide to meet the technological needs of the overseas markets. Meanwhile, the fiercely competitive environment also pushes MNCs to seek more overseas resources, from foreign researchers and information of technological development to foreign R&D capitals and equipment, to meet the technological needs for market competition across the world.

Based on Japan, Saito Yu also categorized the internationalization of the N–R relation into two groups, expressed as the Nd and Rd of home countries and the Nf and Rf of other countries, and specified the models of R&D internationalization. In the first model, the domestic resources (Rd) for technological development are used to meet the needs of technological development overseas (Nd), like undertaking commissioned overseas research to develop new products for export. In the second model, when there are no sufficient domestic resources for technological development (Rd) or when using foreign resources for technological development (Rf) is more favorable, then Rf is used to meet the domestic needs for technological development (Nd), like commissioning foreign countries to carry out research or setting up overseas R&D institutions.

Saito Yu clarified the concepts "needs for technological development" and "resources for technological development", with which he analyzed how MNCs use existing R&D resources to meet overseas and domestic needs and proposed that when there are not sufficient existing resources, MNCs can acquire external resources for technological development through overseas investment to meet the needs for competition during their continuous development. When an enterprise develops a technology, there is first of all an external market demand for the technology. The enterprise then uses existing resources for technological innovation based on the external needs. The accelerating technological transformation and increasing diversified demand for innovation have imposed higher requirements on the innovation of MNCs. However, sometimes the parent companies may face bottlenecks in the supply of resources. The techno-logical innovation, which is complex, comprehensive and risky, makes it inevitable for MNCs to change the original practice of concentrating all the R&D in the parent companies. Besides, as the revolution of informa-tion technology has transformed the traditional ways of R&D, remote computer control and information exchange make international R&D cooperation possible.

1.2.3. *Technological development and augmentation theory*

Kuemmerle (1997) divided FDI in R&D of MNCs into two types, namely, home-based exploiting (HBE) and home-based augmenting (HBA). HBE has been the mainstream understanding over the past few decades, while HBA approaches R&D of MNCs from the perspective of resources. Kuemmerle believed that the major drivers behind R&D internationaliza-tion should include MNCs using overseas resources to expand the home research bases. Since different countries may have R&D resources that are different from those of the home countries, MNCs establish R&D institutions in the countries with target resources and cooperate with the public research institutions, universities or enterprises in host countries, to acquire new technologies and knowledge, increase the stock of knowl-edge of the parent companies and enhance their international competitive-ness. To be specific, HBE is to develop the technological advantage created by the location of the home countries, which is more profitable than other choices. External investment is mainly aimed at making full use of the existing technology and knowledge of MNCs to expand the international market. The market size and potential for growth are the major factors influencing the FDI in R&D of MNCs. HBA is to acquire new technology and knowledge for the home country, increase the stock of technology for the parent companies and improve the international competitiveness of MNCs in the era of knowledge economy, so as to guarantee the stable and long-term growth of MNCs. The investment in R&D of the public institutions and the private organizations, the quality of domestic human resources and the excellent achievements in related fields of the host countries are the major factors that influence MNCs in choosing overseas locations and setting up R&D institutions.

According to Kuemmerle, for HBE FDI in R&D, products' adaptation to location, the diversified production of products and technologies' adaptation to location are emphasized. For HBA FDI in R&D, the creation of core products, the support for core products and the creation of technologies are the highlight.

1.2.4. *Centralization–Decentralization Theory*

Pearce (1989) divided the forces that influence MNCs in setting up R&D institutions into centrifugal force and centripetal force. Centripetal force means the R&D activities of enterprises are centralized in order to control such activities and economies of scale, while centrifugal force refers to foreign R&D investment for R&D internationalization. The balance and adjustment of the two forces make up the centralization–decentralization theory. The motivations for centralization include protecting the exclusive technologies of the enterprises, realizing economies of scale of the R&D activities, reducing the control of R&D and coordinating costs and path dependency. The motives behind the centrifugal force include needs, supply and strategy.

Cheng and Bolon (1993) believed that different conditions, motivations and environmental factors influence the FDI in R&D of MNCs. These conditions include external and internal factors. External factors like progress in information technology facilitate MNCs' coordination and integration for internationalization. Internal factors include new technologies acquired from abroad and support for foreign productive enterprises. Granstrand divided the factors influencing R&D internationalization of

MNCs into driving factors and inhibitory factors. Driving factors are the centrifugal force that decentralizes the R&D institutions of enterprises, including supporting localized production to meet the demand of local consumers, acquiring advanced technologies from other countries, reducing labor cost for R&D as well as the influence from the policies of the host countries and the major foreign acquisition. The inhibitory factors are a kind of centripetal force that makes the research institutions of MNCs centralized and aggregated, including strictly controlling and monitoring the R&D activities, reducing the risks of leaking information of technology, getting close to the domestic market, realizing economies of scale of R&D, lowering costs of coordination and communication as well as influence from the policies of the home countries. When MNCs are making R&D strategies, they have to balance these two kinds of factors.

1.2.5. *Transaction cost theory*

The transaction cost theory is inspired by many scholars' emphasis on the particularity of knowledge of enterprises. Since it is challenging to determine the quality of knowledge products, the sellers often have to explain to and communicate with the buyers during the transaction process, which indicates greater possibility of leaking information of the knowledge products in the process. Creating knowledge is a lot costlier than copying knowledge, which is easier. Since knowledge is so unique as a public good that it is difficult to use it through market transaction, enterprises that create knowledge face the problem of how to benefit from the right of using knowledge. In other words, it is no mean feat for enterprises with some kind of knowledge that may generate monopolistic advantages to negotiate transaction through the external market. Hence, MNCs give up this kind of transaction. When transferring investment capital to other countries, they also transfer the technicians with technical know-how, experience or knowledge, to make it easier to use their ownership assets and provide technical and knowledge support for their overseas operations to improve the adaptation of such operations. In essence, the transaction cost theory takes the R&D internationalization of MNCs as a way to acquire rental for knowledge and such a way has been internalized.

The theory holds that R&D internationalization of MNCs, especially cooperative research, helps deal with the following kinds of transaction cost.

The first category is communication cost. Among the collaboration parties, enterprises tend to have specific yet changing needs. They do not know much of the capabilities of universities and research institutions. Thus choosing partners from them will cost them a lot of time and energy. Universities and research institutions also need to spend a large amount of resources learning about the enterprises and their needs in order to carry out cooperation and innovation. As a special economic behavior, enterprises, universities and research institutions tend to have different value orientations and cultural backgrounds that hinder communication among different parties of cooperation and innovation. A stable R&D partnership cuts the communication cost.

The second category is the negotiation cost. This is related to the positions, negotiation skills and psychology of the parties involved. In industry–university–research collaboration and innovation, the parties have unequal positions. Although the technology sellers (universities and research institutions) possess the technological resources, they need more economic resources. Enterprises need technological resources and they tend to have more suppliers of such resources. Therefore, in the technology market, the buyer market emerges prior to others. Another factor influencing the negotiation cost is values. Generally speaking, what matters more to the intellectuals is that their fruits are applied while enterprises attach more importance to lower transaction cost. A stable R&D partnership cuts the negotiation cost.

The third category is the cost of implementation. For enterprises, there are risks of whether they can obtain the technologies as required by the contract object. For universities and research institutions, there are huge payment risks of whether they can receive all the cooperation funds in time in accordance with the contract. Therefore, all the parties engaged must spend a huge amount of money so as to ensure that the contract is executed. A stable partnership for R&D cuts the cost of implementation.

The last category is risk cost. Although all the parties have duly executed the contract, if the effect of using the technology is not desirable due to changes in the market, the technology still fails to bring the expected benefits to the enterprises and even cause losses. For universities and research institutes, contingent events like job-hopping of the researchers and damage of the lab devices can increase the transaction cost. A stable partnership for R&D cuts the risk cost.

2. Studies on Motivations for the R&D Internationalization of MNCs

Studies on motivations for the R&D internationalization of MNCs have been a key research area for scholars. Early studies believed that MNCs set up foreign R&D institutions in order to make full use of the technological advantages of their parent companies (Krugman, 1981; Lall, 1979). In line with the international product life cycle theory, this conclusion reveals that the mission of the foreign R&D institutions of MNCs is to receive the technology transferred from their parent companies to host countries. Later research, however, pointed out the insufficiency of this R&D centralization theory in explaining the R&D internationalization of MNCs and added another motivation — the advantage of technological resources of host countries (Fors and Zejan, 1996). Dunning (1996) argued that an increasing number of enterprises make foreign R&D investment not merely to expand existing competitive advantages but also to acquire new advantages or complementary assets so as to maintain or enhance their global competitiveness. Mutinelli and Piscitello (1998) believed that MNCs are not able to create internally the capacities to safeguard or augment their global competitive status, which becomes a strong drive behind the foreign R&D deployment. Based on the existing studies, we categorize the theories on the motivations for R&D internationalization of MNCs as follows.

2.1. *Need for technological localization*

One of the major motivations for R&D internationalization of most MNCs is to design products suitable for the market of the host countries through R&D to support their production activities there. MNCs' foreign branches need to carry out R&D in host countries so as to design and develop products and production processes that meet local market conditions. Multinational operations are greatly different due to the differences in cultural backgrounds, customs, consumption behaviors and preferences. R&D internationalization allows MNCs to get closer to the market and consumers of host countries, learn about local demands, and develop products and production processes accordingly. To produce locally products with style and taste that tally with the local consumers' demand, MNCs have to engage in R&D locally. The "global products with a local taste" requires MNCs to take the initiative to research into and develop products with their technologies for the local market.

Since MNCs aim to maximize their profits worldwide, they invest in the production and sales of the overseas subsidiaries and make the home countries set up R&D institutions in host countries in order to transfer the reviewed advanced technologies from the parent companies to the subsidiaries to meet the demand of the local markets. Moreover, after they set up R&D institutions in host countries, they carry out related basic research to develop products for the global market. Thus, the overseas R&D institutions are established through the process from technology transfer by developing products for local market to gradually becoming a technological development center in the global operations. Such a process is in line with the traditional technology life cycle theory.

2.2. *Need to acquire technology from competitors in host countries*

When the technologies of MNCs are not at the frontier of the industry, R&D internationalization may be their way to acquire or track the advanced technologies of host countries and to benefit from the technological spillovers of the R&D of host countries. In this case, R&D internationalization of MNCs serves two functions, namely, innovation and development, as well as absorption of the technology spillovers of other enterprises (Cohen and Levinthal, 1989). Jaffe (1986) believed that, in the technical fields that have attracted many enterprises, MNCs have to spend more on patent, profit or market value in their own R&D and the technological spillovers inside the industry increase as the distance decreases (Jaffe, 1986). Compared with means like obtaining a license, purchasing a patent and hiring the technical professionals from competitors, dependent R&D is the most effective way to "learn" from other companies' products and production processes. Some MNCs set up R&D institutions abroad as an information window and a foothold for their parent companies so as to acquire the latest information worldwide in time. By collecting, collating and processing the technological information, they develop new technologies and products that meet the strategic requirements of their parent companies to eventually deliver the goal of transferring the fruit of technological development back to their home countries. The R&D institutions that many developing countries set up in such developed countries as the United States and Japan are designed to make use of the advanced technologies and R&D environment so as to serve domestic enterprises and R&D institutions of the home countries.

2.3. *Utilizing favorable R&D factors of host countries*

The technological resources available for R&D activities include technical professionals as well as the sound technical environment necessary for successful R&D, such as the technical facilities, laws and policies. The heightening global competition has given birth to an increasing demand for quality R&D professionals. Since it takes a long time to train professionals and their number and quality are restricted by many factors, many countries are facing a shortage of R&D professionals. Such a shortage at home can be addressed when MNCs set up foreign R&D institutions, hire foreign technical professionals and backbone as well as seek R&D talent from across the world. Some MNCs conduct R&D abroad to reduce the cost of R&D. Host countries with abundant cheaper yet well-trained R&D professionals or other resources necessary to the technological activities are attractive to the R&D institutions of MNCs.

Since the 1990s, MNCs have taken over the field of R&D in China, mainly to make use of the abundant yet cheap technological professionals there. China boasts a globally recognized intelligent population. Its technological professionals tend to be equipped with a solid foundation, excellent research qualities and dedication. However, the existing research management system hinders the conversion of technological products into productivity. Many researchers lack the opportunities to give full play to their knowledge and intelligence. Therefore, MNCs can mobilize their proactivity and make use of their talent with input far lower than the international level.

2.4. *Foreign R&D as complementary assets*

American scholar Serape (1999) used the term "complementary assets" coined by Teece (1986) and developed the complementary assets theory for the foreign R&D investment of MNCs. According to Serape, MNCs engage in FDI in R&D to ensure the safety of the key internal assets and help give full play to those assets. For instance, some MNCs set up overseas R&D bases to improve their basic research capacities by accessing the local basic research information. Some improve their domestic basic research through investment in foreign research on application or development. Others improve their production technologies by investing in the R&D of processes and technologies abroad. Still others engage in foreign R&D investment to obtain the key complementary technologies. Serape believed that the FDI in R&D of MNCs is aimed to provide complementary assets that are crucial to the success of MNCs in overseas production and operations. The overseas operations and production transfer of MNCs often entail the adaptive development of products, some of which should be redesigned or need revised processes. Under such circumstances, setting up R&D institutions in foreign locations is necessary in order to support the overseas production and operations. The complementary assets theory also sheds light on why many MNCs set up R&D institutions in different locations abroad and incorporate them into a global integrated research and development network. The key technological assets of products or innovation may be distributed among the research institutions in different locations. Thus, the innovation of advanced technologies is carried out in different research locations abroad at the same time. For maximum profit from R&D of the complementary technologies, MNCs set up overseas R&D branches in the locations with advanced technologies and incorporate them into the same global R&D research network. This is the role of R&D as complementary assets illustrated by Serape from the perspective of technology acquisition.

Besides, as the status of game theory in economics is enhanced, it has been widely applied to explain the real-life social and economic problems. Some scholars also use it to examine the R&D internationalization of MNCs with the view that it is in fact a game involving MNCs, home governments, host governments, host enterprises and other MNCs. The result of the game is that MNCs gradually set up R&D institutions around the world, form strategic R&D alliances and build a global R&D network.

3. Selection of Location for R&D Internationalization of MNCs

Theories on the selection of location for foreign R&D investment of MNCs mainly focus on the various factors that influence such selection. Scholars in the West have carried out a great number of theoretical and empirical studies.

Factors influencing selection of location fall into two categories, external and internal factors. This section is dedicated to the external factors influencing MNCs, i.e., the location factors of host countries.

3.1. *Investment stock of MNCs in host countries*

The investment stock of MNCs in host countries has influenced their selection of location for R&D internationalization in four aspects. First, in terms of the market share of the investment, Odagiri and Yasuda (1996) investigated the foreign R&D of 1,563 Japanese companies and discovered a close connection between the internationalization of MNCs and their market share of host countries. Second, concerning the way to use investment stock, Hakanson (1995) believed that those adopting merger and acquisition are more likely to set up R&D institutions than those making a greenfield investment. The reason is that MNCs tend to keep the R&D institutions of the merged or acquired enterprises, while greenfield investment faces many uncertainties that prevent the R&D activities that serve the productive institutions from being carried out effectively. Third, the nature of industries of the investment stock also matters. According to Lall (1979), foreign investment of MNCs in the process-based industries lags behind that of MNCs in the project-based industries. Pearce (1989) further revealed that, for the process-based MNCs, the end products are produced in home countries, while production in overseas factories is based on the design from the R&D departments back in the home countries. For project-based MNCs, end products must be produced overseas to adapt to the local demand, which makes it imperative to set up overseas R&D institutions. The last aspect is the technological content in the investment stock. Looking into the investment in Europe of 799 Japanese enterprises, Mariani (1999) found that MNCs with moderate technological content have the largest scale of R&D internationalization. Lee and van der Mensbrugghe (2001) found that MNCs with low technological content enjoy a higher level of R&D internationalization, which they believed is because MNCs that engaged in the industries with high technological content are more prudent in overseas expansion due to high confidentiality. However, Patel and Pavitt (1998) argued that high-tech products usually do not need to adapt locally and the level of R&D internationalization is not related to high technology.

3.2. *Accumulation of R&D factors of host countries*

A large number of R&D factors are needed to carry out R&D activities. Hence, the accumulation of R&D factors of host countries is a major determinant of MNCs' selection of location for R&D internationalization. Kumar (2001) believed that rich R&D human resources can effectively attract MNCs. The sound technological resources of host countries may influence the local R&D activities of MNCs and guide them to engage in R&D activities in line with the comparative advantages of host countries. Miller (1993) analyzed the factors that influence the location of R&D institutions of 20 automobile companies in North America, Europe and Asia and found that the R&D institutions are mainly established as "surveillance outposts" to track the projects and design activities of the competitors. In these R&D institutions, engineers are responsible for penetrating the local network, evaluating the trend of technological development and proposing new technical design concepts to the headquarters. Focusing on the overseas R&D institutions of the Swedish MNCs, Fors and Zejan (1996) concluded that they tend to set up R&D institutions in countries that are highly specialized in technologies of their field. Odagiri and Yasuda (1996) examined the overseas R&D activities of 254 Japanese manufacturing companies and believed that the R&D investment in Europe and the United States is targeted at acquiring new technological information. Papanastassiou (1997) pointed out that the technological development of host countries is in a positive correlation to the R&D investment of MNCs. Dalton and Serapio (1999) analyzed the overseas R&D investment of the United States and the expenditure on foreign-funded R&D inside the United States and proved that the United States mainly invests in countries with a high level of technological expertise, while MNCs investing in the United States are attracted by its technological level. According to Meyer and Reger (1999), as advanced R&D resources are mainly available in a few countries, the motivation behind R&D globalization has shifted from adaptive improvement and change in the early days to acquisition of advanced technological information and high-tech knowledge.

In addition to the degree of accumulation of R&D factors, the R&D cost of host countries also determines the selection of R&D location of MNCs. As Bhatnagar *et al.* (2003) revealed, some MMCs select locations in host countries with a less developed economy mainly to reduce cost or get close to the market and in host countries with a more developed economy for their infrastructure, human capital, market structure, industrial connection, policies and environmental factors. Reddy (2000) suggested that MNCs choose to set up R&D institutions in India to cut their cost because the expenditure on research professionals takes up the largest proportion of the overseas R&D cost.

3.3. *The degree of perfection of the market and industries of host countries*

The overseas production scale of MNCs and the market size of the host countries are the most important determinants of the selection of location for R&D internationalization. Mansfield *et al.* (1979) studied 55 American manufacturing companies and found that the ratio of the sales of the overseas subsidiaries to that of MNCs was in a positive correlation to the ratio of the R&D expenditure of the overseas subsidiaries to the overall R&D expenditure, and in a negative correlation to export. Hirshey and Caves (1981) drew a similar conclusion based on the data on 24 American MNCs in 1996. Zejan (1990) investigated the foreign R&D investment of Swiss MNCs and believed that the market size and the per capita income of host countries have a positive influence on MNCs' R&D investment there. It is more likely for overseas subsidiaries engaged in production for markets of other countries to take up R&D activities.

To be specific, first of all, the market size and level of economic development of host countries affect the selection of location. Zejan (1990) and Hakansson (1993) both proved that market size is a decisive factor for the choice of R&D location. Dobson (1997) pointed out that host countries with a large domestic market may attract MNCs to make R&D investment or transfer technology, which constitutes their unique location advantage. Kumar's (2001) analysis on the selection of location for overseas R&D of the American and Japanese MNCs showed that the domestic market size and the degree of participation in the regional trade alliances of host countries are the major reasons why MNCs set up local institutions. The market structure of host countries is another factor. Saxcé (1994) and Cantwell and Santangelo (1999) proved that the market structure of host countries also has a major impact on the R&D investment decisions of MNCs. Cantwell *et al.* (2001) asserted that MNCs do not set up R&D institutions in host countries whose market is not competitive enough, but in a market structure with monopolistic competition, MNCs compete to set up R&D enterprises out of strategic competition. The last factor is the industrial connection between host countries and MNCs. In their study on the selection of location for overseas R&D of the Swedish MNCs, Fors and Zejan (1996) concluded that the industrial comparative advantages of host countries or the industrial connections between host countries and MNCs are the major factor. The overseas locations pre-ferred by the manufacturers usually enjoy specific R&D advantages in line with the technical fields of the manufacturers.

3.4. *The economic conditions of host countries*

Since decision-making for R&D internationalization is a means for MNCs to maximize profit worldwide, they take into consideration the cost and benefit of R&D investment when deciding the locations. Such cost and benefit are affected by the economic conditions of host countries. The economic environment of host countries, including the infrastructure (communication, transportation, electricity, etc.), policy orientation and geographical environment, affects MNCs' selection of location for R&D investment. Kuemmerle (1997) discovered that the selection of location for overseas R&D of the American MNCs is affected by factors like the scale of production of the subsidiaries, the business nature and the techni-cal resources of host countries, as well as their policy environment, espe-cially the protection of intellectual property rights. Kumar (1995), however, drew a different conclusion that MNCs do not conduct R&D for technologies in need of protection in other countries and thus the protec-tion of intellectual property rights has no major impact on MNCs' deci-sion on overseas R&D. Kumar (1996) pointed out that communication facilities are the fundamental condition that attracts MNCs. Niosi (1997) believed that host countries provide a large number of preferential policies to attract investment from MNCs and help them lower the cost on foreign investment. Ernst and Palmer (1998) noted that the South Korean govern-ment offers tax deduction on the reserve fund for technological develop-ment and the R&D expenditure and accelerated depreciation for the R&D facilities of foreign investors, which has effectively promoted R&D investment from European and American MNCs in South Korea. In addition, the trade policies of host countries are also a determinant of MNCs' selection of location for R&D investment. On the one hand, the restrictions on import of intermediate products prompt MNCs to carry out technological development in host countries. On the other hand, extremely strict import restrictions undermine their proactivity in R&D investment (Kumar, 1996).

3.5.   *Institutional and cultural background of host countries*

When MNCs are deciding the location for R&D internationalization, the policy and institutional environment of host countries are the important guarantees that decide the risk and the benefit of R&D investment. After China's entry to the WTO, the institutional reform of China is a decisive factor that drives MNCs to make R&D investment in the country. Through empirical study on the American enterprises' selection of location for overseas R&D, Doh *et al.* (2002) discovered that MNCs also take into consideration the institutional factors in their choice of location, in addition to market size and technological development. Among the institutional factors, the level of protection of intellectual property rights has been a crucial component. However, the conclusions of related studies diverge. On the one hand, a country's level of protection of intellectual property rights seems to be the decisive factor behind the R&D activities of MNCs which do not set up key R&D institutions in countries that lack such protection areas. On the other hand, if MNCs are engaged in technology-based or locally adaptive overseas R&D activities, it is relatively easier to control knowledge. In this case, the condition of protection of intellectual property rights does not have a major influence on the selection of R&D location.

In addition to policies, laws and other formal systems, culture, a component of the informal system, also affects the decision of MNCs for R&D internationalization. Hakanson (1995) believed that a shared professional background cannot reduce the cultural differences among researchers of different nationalities in the R&D activities of MNCs. Hofstede (1980) asserted that different cultural backgrounds have different influences on corporate innovation. Countries with a better incentive culture for innovation are more likely to attract R&D investment from MNCs. Laamanen (2002) studied three transnational R&D merger and acquisition cases in Europe and revealed the influence of culture on the integration of R&D institutions after the merger. Kedia *et al.* (1992) pointed out that MNCs should mainly locate their R&D institutions in regions where the cultures help boost R&D efficiency.

3.6. *Selection of location inside host countries for R&D internationalization*

For R&D internationalization, if there are regional differences inside the market of host countries, or there is apparent market segmentation, the selection of location inside host countries becomes a more essential problem. However, research in this regard tends to be ignored by scholars. Our case study of China reveals that, no matter which host country MNCs are in, the distribution of R&D investment is extremely uneven and is becoming increasingly intensive. In China, most of the R&D institutions of MNCs are located in Beijing and Shanghai. On the whole, MNCs share the following features inside host countries.

Close to the regional headquarters of MNCs: As a major component of corporate function, R&D mainly serves the production and sales of a company, works closely with the departments of production and sales, and offers the R&D researchers more opportunities to access the market. Such a connection will be undermined if the R&D institutions are far from the headquarters. Besides, the management of the company will prefer to have direct control over the R&D institutions. If they are close to the headquarters, there will be more opportunities for communication between the two.

Near universities or research institutions: Universities and research institutions gather most of the technological resources of host countries. For MNCs, especially those seeking resources, setting up R&D institutions near universities or research institutions, first of all, allows them to make full use of the spillover effects of those innovative organizations to serve their technological innovation. Additionally, it is easier for them to cooperate with those innovative organizations in R&D. Moreover, they may find suitable R&D professionals from the universities and research institutions, which gather a lot of talent.

Located in economically developed core cities: Big cities are the hubs of political, economic, cultural and technical activities of host countries, and are thus most preferred by MNCs as the location for R&D institutions. Only big cities can offer better conditions for R&D activities. They also set the trends of consumption. Hence, they are the first choice for MNCs to learn about the market of the host countries.

Concentrated in the high-tech parks: High-tech parks enjoy evident advantages in attracting R&D investment. First, intensive in knowledge, talent and technology, they bring together enterprises, universities and research institutions and give full play to the agglomeration effect of innovation. Second, R&D institutions enjoy preferential policies in the high-tech parks that allow them to reduce the overall R&D costs and improve R&D efficiency.

3.7. *Related studies in China*

Chinese scholars did not pay attention to the selection of location for R&D internationalization of MNCs until the late 20th century and early 21st century. Zhang (1998) first studied R&D internationalization from the perspective of strategy of MNCs. Based on the strategic goals, she divided R&D institutions into two modes, namely, home-based growing mode and home-based utilizing mode. The former is mainly found in areas where most research institutions and universities gather, while the latter is found in areas with a large market size. Lin and Chai (1998) pointed out that the proportion of foreign R&D investment of MNCs to the total R&D fund keeps growing, the proportion of R&D professionals in the foreign enterprises has greatly increased and cross investment among high-tech fields has been significant. Besides, MNCs tend to develop technologies in host countries with less strict research policies, better service facilities and huge market potential. Du (2005) divided R&D institutions into three types based on the differences in motivation of investment and the selection of the location. R&D institutions that offer production support tend to be located near overseas production bases in countries or regions with a large market size. R&D institutions that track technologies tend to be found near competitors in countries or regions with a high level of technological development. R&D institutions that utilize talent and resources tend to be situated in technological core cities of countries and regions that boast of abundant talent resources and a sound technological environment, especially near famous universities and research institutions. However, he simply put infrastructure, laws and policies under the category "resources". In fact, infrastructure, laws and policies only make up the environment for investment. The environmental factors should be separated from the resource ones.

4. Spillover Effects of R&D Internationalization of MNCs

The technological diffusion and spillover effects of MNCs are widely discussed topics across the globe. Scholars have shown great interest in it and have analyzed it with different research methods, perspectives, as well as data collection and processing methods. As a result, the conclusions vary. The R&D internationalization of MNCs has a dual influence on the technology of host countries. On the one hand, MNCs transfer technology directly. On the other hand, they influence the technological capacity of host countries through technological spillovers (Blomstrom and Kokko, 1996; Baldwin *et al.*, 2000). This section focuses on the technological spillover of the R&D internationalization of MNCs. Technology transfer will be discussed in the next section.

Technological spillover of R&D internationalization means that, during the R&D of MNCs in host countries, technologies involuntarily diffuse in host countries and lead to their improvement in technology and productivity. MNCs set up R&D institutions in host countries either to serve their global production and market network or to be a component of their global R&D network. Meanwhile, enterprises that have business connections with the branches of MNCs are also indirectly included in the networks. As the network effect and synergy increase, learning and innovation become increasingly intensive (Nadvi *et al.*, 1995; Ernst, 1998), which leads to closer technical cooperation and sharing between MNCs and host countries. There are four major spillover channels.

4.1. *Linkage effect of human capital*

The linkage effect of human capital refers to the flow of technology, management and other important professionals between the R&D institutions of domestic enterprises and those of MNCs (Jiang, 2004). The flow of human capital is one of the reasons why the technology diffusion of R&D internationalization of MNCs and the spillovers thereof lead to technological progress of host countries. Lall (1979) pointed out that linkage is the mutual influence between MNCs and the domestic enterprises beyond the "pure" market transaction. He divided the linkage between MNCs and enterprises of host countries into three types, namely, backward linkage, forward linkage and horizontal linkage. Backward linkage emerges when

MNCs sell products to the enterprises of host countries. Forward linkage is generated when MNCs purchase products from the enterprises of host countries. Horizontal linkage is the competition effect between MNCs and the enterprises of host countries. Caves (1982) found that the flow of corporate managers from Japan to the United States and Europe speeds up the spread of specialized managerial skills. The flow here has several connotations, including the tangible transfer as well as the intangible transfer of human capital (Zheng, 2006).

MNCs' training of local employees is the foundation of the indigenous innovation of host countries. This kind of training is open to all the levels. From the simplest production operators to supervisors, from senior technicians to upper management, almost every level has the opportunity to receive training. There are various ways of training, including on-site guidance, expert seminars and even overseas formal education. Apparently, when an employee moves from an MNC subsidiary to another enterprise or an indigenous innovation company, the various skills that he has learned in the subsidiary also flow outward. In this way, the influence of FDI on indigenous innovation is at work.

4.2. Demonstration effect of technological R&D

The demonstration effect of technological R&D includes four aspects: The first is the demonstration effect of new products and technologies produced by MNCs' R&D internationalization on the enterprises of host countries, namely, the demonstration effect of products and technologies. The second is the demonstration effect of the technology management model in MNCs' R&D internationalization on the enterprises of host countries, i.e., the management demonstration effect. The third is the demonstration effect of the mode and orientation of developing new products and technologies of MNCs' R&D internationalization on the enterprises of host countries, i.e., the product development orientation effect. The fourth is the demonstration effect generated when some MNCs are willing to have multifaceted technical exchanges with enterprises of host countries, namely, the technology exchange effect (Jiang, 2004).

Due to the technological gap between MNCs and local enterprises of the host countries, the latter can bring about indigenous innovation by learning and imitating the former's technological innovation behaviors. This kind of effect involves two processes. One is that MNCs enable subsidiaries to acquire international advanced technologies through R&D internationalization and the other is the subsidiaries of MNCs diffuse the technologies they acquire to the economies of the host countries. For a rational economic entity, given the dominant technological monopoly it has obtained, it has no obligation and there is no necessity to diffuse technologies to host countries. Therefore, two types of technology diffusion emerge, namely, indigenous innovation brought about by voluntary technology diffusion and that by involuntary technology diffusion. The former is technology transfer realized by cooperation and other means between MNCs and the enterprises of host countries in the R&D internationalization of MNCs (Schrader, 1991). The latter is the technology spillover effect highlighted in this section and may be considered mainly as the technological progress achieved in two ways, namely, reverse engineering and the imitation behavior of the enterprises of host countries (Swan, 1973; Riedel, 1975) as well as the flow of human capital (Mansfield, 1980). When the economic development level of host countries is not high, technological progress brought about by voluntary technology diffusion dominates (He, 2000).

Lall (1999) summarized 10 characteristics of host countries' technological learning, i.e., the acquisition of demonstration effect. The characteristics are the importance of learning technology, incompleteness of information about enterprises of alternative technologies, uncertainty of the learning process, path dependence of the learning process, non-transferability of learning ability, specificity of technological spillover effect, locality of R&D innovation, incremental risk at technical level, eternality and relevance of technological learning, and transnationality of technological interaction.

The R&D internationalization of MNCs allows display of their advanced products, processes and management methods. This kind of display has a demonstration effect on local enterprises and exposes local enterprises to advanced foreign technologies and level of management by opening up. To compete with MNCs, local enterprises often try to imitate their technologies. They can employ their own technical strength to acquire MNCs' technology through imitation, introduce, assimilate and absorb technologies, or employ former employees of MNCs for imitation to improve their own technical level. Another way for manufacturers to acquire technical knowledge is "reverse engineering", which means studying products to obtain their production technologies. Reverse engineering is an important means of imitation.

4.3. *Competition effect of product market*

Competition effect means that, through R&D internationalization, MNCs break the original monopoly pattern as new competitors and force local enterprises to accelerate the development of technology and products. On the one hand, competition effect is the pressure from MNCs' R&D investment on the enterprises of host countries. It forces them to increase their R&D investment and improve their R&D efficiency to prevent market competitiveness from being weakened. On the other hand, the R&D investment of MNCs in developed countries is also likely to generate two-way spillovers. In other words, the R&D institutions of MNCs may have technological spillovers on enterprises of developed host countries and receive reverse spillovers from the R&D activities of the enterprises of the developed host countries (Patel and Pavitt, 1993; Reger, 2001). Such reverse spillovers can be regarded as a kind of competition for host countries, the effect of which should be more complicated. When MNCs obtain market and technological monopolistic advantages in host countries with their advanced international technologies, competition effect on the enterprises of host countries emerges. This kind of effect mostly appears inside the industry that MNCs belong to and is thus technological spillover inside the industry (Shung *et al.*, 2015). For different industries, the competition effect takes distinctively different forms. For the weaker national industries or the more competitive industries of host countries, the entry of FDI will compete for the resources of the original market and generate competitive pressure on the enterprises of host countries. Under such circumstances, the enterprises of host countries will without doubt increase R&D investment to redistribute market resources, thus promoting improvement in indigenous innovation capacity. For industries with a high degree of monopoly in host countries, the entry of FDI breaks the original monopoly pattern and the welfare loss caused by monopoly will be compensated. In other words, the welfare of the entire industry will increase. As a region with a higher level of economic development in China, Eastern China features stronger industrial competitiveness. Therefore, compared with the demonstration effect, the competition effect brings about indigenous innovation that is more beneficial to the improvement of regional indigenous innovation capacity.

4.4. *Agglomeration effect of R&D location*

Some scholars believe that the agglomeration effect is exerted mainly because of the spatial diffusion of knowledge spillovers, and geographical proximity greatly reduces the cost of knowledge circulation; knowledge exchange among heterogeneous manufacturers within the cluster contributes to innovation (Baptista and Swann, 1998). The location with the combined advantages of R&D factors has a stronger appeal to high-tech industries. Such R&D factors include R&D human resources, technical information, network, infrastructure, work and living environment, as well as social and cultural environment.

MNCs set up R&D institutions in the areas with concentrated technological innovation in host countries in order to establish connection with local R&D institutions and obtain technology spillovers. Once MNCs' R&D institutions are integrated into the network, they themselves reinforce the innovative environment and network within the spatial domain and positively enhance the R&D agglomeration effect. Agglomeration effect is crucial to MNCs' decision-making for overseas R&D investment. In general, MNCs tend to establish new R&D institutions in locations that enjoy a good environment for technological innovation and gather more technological innovation enterprises. Neven and Siotis (1995) believed that Japanese enterprises entered Europe and the United States in the 1960s and 1970s to gain access to the "knowledge source" of competitors. In order to give full play to the agglomeration effect of MNCs' R&D institutions, host countries sometimes set up science and technology parks to provide various preferential conditions and supporting measures for foreign companies to carry out R&D activities, so as to strengthen the R&D agglomeration in the region (Löfsten and Lindelöf, 2002).

5. Technology Transfer System for R&D Internationalization of MNCs

5.1. *Technology trade of MNCs*

Both the contemporary theories of R&D internationalization of MNCs in the context of globalization and the classic theories on the motivations behind R&D investment of MNCs to some extent explain why MNCs internationalize their R&D activities, their methods to reduce risks at the strategic level to better protect their own interests and their ways of technical contact with host countries. However, they fail to pay attention to the simplest economic behavioral decisions made by MNCs as a main body of economic operations.

In fact, R&D internationalization of MNCs is conducted to further realize the differentiation of technologies and firmly grasp the core technologies. Thus, it obeys and serves the "headquarters economy". Based on the curve of "marginal benefits of technologies" and the "maximization of welfare" of international trade in technology, MNCs primarily consider how to obey and serve the headquarters economy in their investment decision-making for R&D internationalization.

In analyzing the flow of technological factors in international markets, it is necessary to first analyze which factors determine the output level of a country. Assume that capital and labor are constant. The output level of each country will depend on the variable technology T and the relationship between the output Q of a country and the supply of factors, i.e., the production function $Y(K, L)$ of a country, as shown in Figure 1.

The slope of the production function Y curve is called the marginal product of technology MPT, which indicates the effect of a small increase in the amount of technology supply on output. As can be seen in Figure 1, when the supply of technology goes up, the total output Y keeps increasing yet at an increasingly slower speed. In other words, the marginal production of technology is declining. In fact, since capital and labor are constant, as the supply of technology continues to increase, the

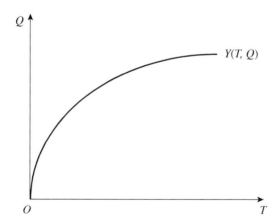

Figure 1.   Production Function of MNCs in the Home Countries.

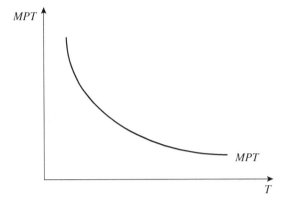

Figure 2. Curve of Marginal Product of Technology.

contribution value of technological factors becomes even smaller, which makes the output of technological increment gradually decline. Figure 2 shows how the marginal product of technology relies on the amount of technological supply.

Now assume that the home country of an MNC *H*s and the host country *F* have different technologies but the same amount of labor and capital supply. Usually, the MNC will first choose to carry out R&D activities in their home country *H*, so that the technological supply in *H* is relatively abundant but the efficiency in resource allocation of domestic technologies is lower. In the foreign country *F*, due to the relative shortage of technological supply, the efficiency in technological allocation is greater. In fact, such a condition causes the international flow of the production factor technology. If the MNC is allowed to shift technological supply freely from *H* to *F*, the technological supply in *H* will decrease, and the marginal product of the technology will go up accordingly. When the technological supply flows into *F*, the marginal output of technology will decline with the increase in supply of technology. If there is no limit to the flow of technology on both sides, the process will continue until the marginal output of technology of the two countries is equal, as shown in Figure 3.

It can be seen from Figure 3 that before the free flow of technological factors, the MNC is engaged in production in *H*, its technological input starting from the origin point on the left is *OT* and the total output is *OHGT*. The technological income is *OBGT* and *HBG* is income from other factors (labor and capital). *OB* is the marginal product of

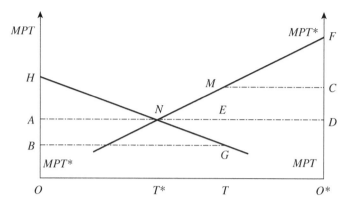

Figure 3.   Economic Analysis of Transnational Flow of MNC Technological Elements.

technology. In *F*, the amount of input of technological factors is *O*T* and the total output is *O*FMT*. *O*CMT* is the technological income and *FMC* is the income from other factors (labor and capital). The marginal product of technology is *O*C*. When technological factors can flow freely, the MNC has technological cooperation with enterprises in F or turns them into their subsidiaries. In this case, the technological factors flow from *H*, which has a lower rate of return, to *F*, which has a higher rate of return. Technological factors flow, move from *T* to *T** and reach equilibrium at point N. Then, the marginal benefit of technology in *HOA* is equal to that in *F O*D*, i.e., *OA* = *O*D*. The total output of *H* is *OHNT** and that of *F* is *O*FNT*. The total output equals net increasing *MNG*.

Next, the change in welfare after the international flow of technological factors and its connotation will be analyzed. As the figures above reveal, for the headquarters of MNCs in *H*, the total income is *OTENH*, the technological income is *OTEA* and *NEG* is the newly added income. For the subsidiaries of MNCs in foreign country *E*, the total income is *O*FNET* and the technological income is *O*DET*. Therefore, in terms of technological factor, the benefits for both the headquarters and the subsidiaries of MNCs increase and the welfare of MNCs rises. Income from other factors of production (labor and capital) falls from *HBG* to *HAN* for *H* and increases from *FMC* to *FND* for subsidiaries in *F*. This suggests that, driven by the benefits of technological factors, the headquarters evolve from other factors intensive to technology intensive and the subsidiaries also become technology intensive, which increases the utility of technology as a factor of production. Therefore, the allocation of technological resources by the MNC in the international market not only increases the rate of return on technology but also reduces the risk of technological investment for their headquarters. The foreign subsidiaries absorb the technological inflow and introduce a large number of new technologies that increase the overall productivity. The increase in output of technological factors of the MNC, a complete system, is the driving force behind its R&D.

5.2. *Analysis of the system of technology transfer*

Competition among enterprises grows increasingly fierce with the deepening of economic globalization. In the international market, if enterprises want to develop a foothold, they must rely on core competitiveness, whose development mainly depends on R&D activities. Through R&D activities, enterprises can master a range of proprietary knowledge, from production processes to technical standards, as well as organizational knowledge in terms of marketing, financing and management. Enterprises achieve the goal of improving production efficiency by virtue of proprietary knowledge and organizational knowledge. Some even transfer technology through patent of technological invention and other means. MNCs rely on leading R&D output to develop and guide industrial product standards and enjoy the benefits therefrom. This is commonly referred to as "a first-rate company selling standards, a second-tier company selling technologies, and a third-tier company selling products".

However, the characteristics of R&D knowledge as quasi-public products affect corporate activities from another perspective. MNCs promote their international technology transfer in various ways in the process of global operation and development. With R&D internationalization, in particular, MNCs begin to carry out R&D activities directly in host countries. This falls within the network paradigm. R&D then becomes the focus of technology transfer. These R&D innovation activities have both technological output effects and technological spillover effects on host countries as well as inevitable technological crowding out effect on host countries. Studies have shown that when host countries' economic system is open enough, they can benefit from the international R&D investment that has integrated international trade, FDI, technology transfer as well as transnational personnel and cultural exchanges. This has to a certain extent promoted the technological progress of host countries, which is obviously not what MNCs have expected. In fact, an in-depth study of the system of technology transfer helps to disclose how MNCs internalize R&D knowledge through rigorous system operations and thereby achieve the goal of protecting their strategic interests.

5.2.1. *Technological innovation process*

The western academic community has paid close attention to knowledge innovation. However, influenced by traditional ideas, scholars tend to think that organization is an information processing system (Simon, 1973) and knowledge innovation is only a part of information processing, so few have delved into the dynamic process of knowledge innovation (Nonaka, 2000). In fact, the core competitiveness of technological innovation organizations represented by MNCs is knowledge-based technological capabilities. MNCs realize static reserve and dynamic accumulation of technology and knowledge through organizational and individual learning. With a highly centralized R&D knowledge base, they continuously improve their technical support capabilities to ensure their competitive advantage at the technological level.[1]

Ikujiro Nonaka (1995) first proposed the SECI model in *The Knowledge Creating Company*, thoroughly analyzed the knowledge creation process and comprehensively discussed the knowledge field of knowledge assets and knowledge innovation. Rao (2003) believed that the SECI model accurately specifies the starting and ending points of knowledge production at a theoretical level, clearly distinguishes the main categories of knowledge production models and creatively proposes a tool for assessing enterprises' performance in knowledge management. However, the SECI model indicates that the process of generating new knowledge, i.e., knowledge creation, is equivalent to knowledge innovation. "Innovation" is not strictly different from "creation". In fact, knowledge innovation has a broader denotative meaning. Knowledge creation is embedded in knowledge innovation. In addition, influenced by Japanese philosophy and practice, he mainly studied general knowledge in the process of knowledge creation but ignored the importance of explicit knowledge in organization.

---

[1]Rui, M. J. *et al.* Research on Knowledge Innovation Models of High-tech Enterprises. www.fdms.fudan.edu.cn/teach-erhome/rmj.

Rui (2005) used the Holsapple (1998) concept, "system knowledge value chain", to decompose the enterprises' knowledge value chain into five parts, namely, knowledge acquisition, knowledge fusion, knowledge creation, knowledge protection and knowledge diffusion. Meanwhile, the knowledge sharing process is introduced into the same or different levels of enterprises, so that knowledge innovation forms a dynamic knowledge value chain with a feedback mechanism. The dynamic value chain is integrated with three basic modules, namely, the knowledge field of the SECI knowledge creation model, the knowledge creation process and knowledge assets, and derives the enterprise knowledge innovation model, as shown in Figure 4.

As the model indicates, the knowledge innovation process of enterprises includes obtaining new external knowledge through search, purchase, recruitment and learning, selecting and further integrating external knowledge, carrying out the knowledge creation process suggested in the SECI model and diffusing knowledge at various levels of the enterprises to achieve the goal of sharing knowledge. For MNCs, R&D is carried out according to the above process, guides the other parts of the knowledge value chain and realizes the strategic parts of innovating and sharing R&D knowledge.

MNCs acquire R&D knowledge through innovation activities. At the knowledge system level, this includes basic research (including "scientific" components), proprietary knowledge, production processes and know-how, and therefore has rich connotations. The R&D knowledge is mainly manifested as the "proprietary technologies" that are owned by

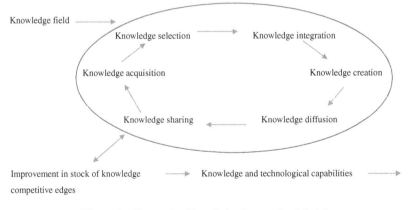

Figure 4.   Enterprise Knowledge Innovation Model.

MNCs and directly serve production and operation. However, these kinds of proprietary technologies are generated by the knowledge creation activities of MNCs. It is a special commodity with a price created by MNCs that is different from material products. The proprietary technologies that MNCs exhibit can also be traded as a commodity. However, they have very significant public characteristics, which are summarized by Wu (2000) as follows.

(1) It is difficult for producers of knowledge products to control the results of knowledge innovation. If the producers hide the knowledge product, their innovation activities will not be recognized by the society and thus become meaningless. However, if the producers disclose the knowledge products to the public, they cannot effectively control information, the intangible resources.

(2) The personal consumption of knowledge products does not affect the consumption of other individuals and countless individuals can share open information resources. Intangible knowledge products are disclosed in the form of tangible carriers and have the "publicity" in the economic sense.

(3) Knowledge products are a kind of perishable asset. Although the production of information comes at a cost, the cost of transmitting information is relatively small. Once the information of the producers is obtained by the competitors for free or at a price lower than the cost of production, the competitors will become "free riders" of the information.

These three characteristics mean that the R&D knowledge acquired by MNCs in R&D activities will also bring spillover benefits to rivals, which indirectly enhances the strength of the rivals. This phenomenon is particularly prominent in the industrial chain. In other words, the benefits generated by R&D activities of MNCs for themselves are far less than the social benefits, because R&D knowledge has positive externalities which will eventually lead to market failure.

5.2.2. Basic elements of the system

A technology transfer system is an organic whole of a set of interacting and interconnected elements organized according to certain rules. The basic elements of the system include the subject of transfer, the object of transfer and the behavior of transfer.

Subject of transfer: The subject of transfer refers to the people or legal entities with the ability to use technology and make technological innovation, including technology supplier, technology recipient and governments.

The technology supplier in this section refers to MNCs. Their behavioral motivations and their control and capacity of technology transfer have an essential impact on the smooth transfer of technology. Direct investment in transfer techniques can occur when the technology owners' technological capabilities are stronger than those of the technology recipients and technology transfer is a benefit to both sides.

The technology recipients are at the other end of the technology transfer system. They may be the non-associated competitors or partners of the technology suppliers. They may be a business group or other forms of individuals and organizations. A technology recipient must have two major features. First, they must be willing to acquire technologies from MNCs. Second, they should be able to absorb relevant technologies and such ability depends on their existing technological base.

The governments in this section refer to the governments of host countries. It is generally believed that the governments influence the R&D internationalization of MNCs by formulating macro policies. This can be regarded as an environmental factor. However, when the governments directly intervene in the technology transfer process, especially in some major joint-venture projects, then the boundary of the technology transfer system of R&D internationalization of MNCs becomes relatively vague.

Technology intermediaries are an indispensable part of the technology transfer system. Their major functions include providing platforms for technical exchanges (information of technical standards, technical patents, R&D projects, technology trade policies of different countries, etc.) as well as promoting cooperation between the supply side and demand side of technologies (technology assessment, recommendation of technological professionals, technology trade, etc.)

Object of transfer: The object of transfer refers to the technologies, i.e., the R&D results of MNCs as well as the core of the technology transfer system. In this section, the object of transfer is mainly divided into three types based on the technological form. The first type is what we usually call the hardware, which is the physical technological products, such as machines, equipment and reaction devices. In other words, it is the technology reflected in the products. The second type is technologies that exist in written or electronic forms, such as process documents, computer programs and design specifications. They are soft technologies that are organized and accessed in a perceptible manner. The third type exists in the minds of those who master technologies (technical experts), as expressed in their technical experience, managerial skills and technical know-how. In other words, they are intellectual technologies. In addition, technologies also include market development, quality management and other content, which mainly belong to the second and third types.

Behavior of transfer: As long as technological supply and demand exist at the same time, once the subject of technology transfer is determined, the transfer behavior will occur. Two major issues are discussed here, namely, the channels of technology transfer and the barriers to technology transfer.

In terms of how MNCs conduct R&D in host countries, the channels of technology transfer can be simply divided into joint ventures, cooperation and sole proprietorship. Different forms of transfer channels have different effects on technology transfer. A joint venture refers to a jointly owned organization set up with contributions from both parties. It is characterized by joint investment, management and operation. Profits and risks are distributed according to the proportion of registered capital. Cooperation means establishing a contractual jointly owned organization. The two sides share income and fulfill their obligations according to the contract they sign. Sole proprietorship is an organization that an MNC sets up with sole investment in the host country according to the relevant laws. The MNC undertakes all the risks and enjoys all the profits alone. It manages the organization independently.

The barriers to technology transfer are a kind of intangible but objective element that exists through interactions with other elements. The existence of barriers makes it difficult to realize technology transfer, because such barriers generally cause two contradictions: (a) MNCs regard technologies as commodities with a shorter life cycle. In order to recover R&D costs as soon as possible, they usually set a higher transfer price. However, the technology recipients usually regard the transferred technologies as expired commodities and hope to bring down the price as much as possible. (b) MNCs usually control or even monopolize technologies as much as possible, especially for advanced technologies, while host countries aim to introduce and learn advanced foreign technologies. With the above elements integrated, the model structure of the technology transfer system of MNCs' R&D internationalization can be outlined, as shown in Figure 5.

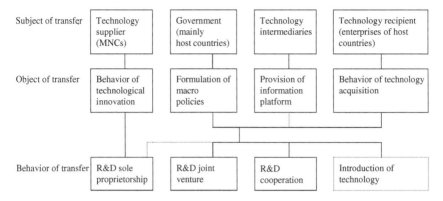

Figure 5. Model Structure of Technology Transfer System of MNCs' R&D Internationalization.

Based on the above analysis, the technology transfer system of MNCs' R&D internationalization has the following characteristics.

The first one is the hierarchy of the system. The system can be divided into several levels, such as a sole proprietorship system and a joint-venture system. This feature allows us to make decisions based on the hierarchical systems and their respective specificities for specific monitoring, reasonable adjustment and effective management.

The second characteristic is the relativity of the elements. The relativity of elements is determined by the hierarchy of the system structure, which helps us avoid a simplified and absolute perception of things. In fact, the technology transfer system of MNCs' R&D internationalization is closely linked to the international economy as well as the national and regional economic systems.

The third characteristic is the ambiguity of the boundaries. The boundaries of the system are relatively vague due to the uncertainty of some technical subjects. For example, in terms of the role and function of technology intermediaries and governments in the technology system, whether they are important cooperators or major promoters depends on the differences in time sequence as well as the differences between countries and regions.

The fourth characteristic is the dynamic stability of the structure. The precondition for the existence of the system is its stability. The internal elements of the technology transfer system of MNCs' R&D internationalization often have a certain degree of interaction with the environment, which renders that the system structure unbalanced. In other words, the structure is dynamically stable.

5.2.3. *Conditions for technology transfer*

Whether or not local enterprises of host countries can benefit from the international technology transfer caused by the R&D internationalization of MNCs depends on the conditions under which technology transfer occurs. From the perspective of microeconomics, the prerequisite for technology transfer lies in the existence of supply of technologies and demand for them. When there is a technological gap between local enterprises of host countries and MNCs, there is, objectively, supply of technologies and demand for them. Furthermore, a gap in the level of technology between the local enterprises of host countries and MNCs should exist and be maintained. MNCs need to conduct R&D activities continuously to ensure the progress and update of their technologies. Hence, technological progress is an essential condition for technology transfer. In terms of technological demand, that MNCs take the lead in technologies and use them to obtain a lot of excess profits will stimulate the technological demand of the local enterprises of host countries. Therefore, the two basic conditions for technology transfer are technological progress and technological demand. Through interplay, they together spur the occurrence and development of technology transfer.

Technological progress has driven the evolution of human production methods. On the one hand, technology follows the characteristics of its own system and continues to advance according to its free development inertia. On the other hand, it is manifested as continuous application of technological innovation. The economic and social benefits generated by technological achievements through commercialization and industrialization in return can effectively stimulate technological innovation and even technological revolution. This thereby enriches the content of the technological system, further accelerates the development of productivity, promotes economic growth and continuously drives economic internationalization, including the internationalization of production, the organization of enterprises and capital. The inner logic of the promotion by technological progress is as follows. As technological progress can create new products, enrich and develop technological systems, it can have a major impact on the development of productivity, economic growth and economic internationalization, as shown in its role as a driving force. Without technological progress, it is impossible to produce a series of new technological products. Without such products, there will be no development of the technological system. Without the development of technology, naturally there will be no improvement in productivity. Without such improvement, there will be no economic growth, economic internationalization or technology transfer. Therefore, technological progress is a prerequisite for technology transfer.

Technological demand affects technology transfer, as evidenced mainly by the demand-oriented and restricted functions. With the market as the medium, it limits and guides the direction of technological development through information transmission. The technologies have value and use value, which is mainly manifested as utility during the process of application, i.e., a direct reflection of the use value of technologies. Based on the contradiction between the limited socially available resources and the continuous expansion of the scale of social demand, it becomes decisive that the growth of GDP and scale growth as well as the improvement in demand structure can only be achieved through technological progress. With information transmission as the chain, technological demand then influences technology transfer by virtue of the role of market mechanism. It begins with the needs and desires of those demanding technologies and finally gets to the technological innovators through a change of market, leading to their behavior of innovation. Therefore, the innovators form a "satisfied response", generate innovative behaviors, create technological products, and then transform them into technological goods through the market. After the behavior of technology transfer occurs, the technological goods successfully reach the hands of the demanders to satisfy their needs. In this way, a technology transfer is completed. Of course, after new demand is generated according to this path, new technological innovation is carried out accordingly and new behavior of technology transfer is finally completed. The cycle is thus repeated.

5.3. *Behavioral decision-making for investment*

Along with the motivations for R&D internationalization investment comes the manner in which MNCs make decisions for R&D internationalization. To deliver the established investment objectives, MNCs must consider which host countries to choose for R&D internationalization, i.e., the selection of location, as well as what forms of R&D to use. What is the final revenue function for investment in a certain host country? These make up the content discussed in this section.

5.3.1. *Motivation and purpose of investment*

A lot of research has been dedicated to the purpose and motivation of R&D internationalization of MNCs. In this book, the major purposes are as follows.

First, MNCs want to take up the local market. Setting up overseas R&D institutions offers direct access to the information on local market demand, which helps develop new products and processes that meet the needs of the host countries' markets. The institutions can also gain insights into their competitors and provide competitive processes, technologies and products in time for the production of MNCs in host countries, so as to occupy the local markets. R&D enables MNCs to carry out local product development and technological research that are characterized by accurate positioning, short cycle and fast response, which greatly enhance the competitiveness of MNCs in the host country markets.

Second, MNCs are looking for R&D professionals. R&D professionals are the foundation of enterprise competition. With increasingly fierce international competition, the demand for high-quality R&D professionals has also surged. However, the training of professionals takes a long period of time and the quantity and quality of supply are restricted by many factors. The shortage of R&D talent is undoubtedly one of the major issues confronting many MNCs. By setting up overseas R&D institutions, they can hire qualified local R&D professionals in host countries, which effectively addresses the above problems and fundamentally maintains and enhances their core competitiveness.

Third, MNCs are building a global R&D network. Forming a global R&D network, first of all, can achieve seamless research. The network connects the whole world, so that R&D institutions across the globe can exchange the research results of the day through the network. This improves R&D efficiency. Besides, R&D resources can be shared globally. The latest research dynamics in the world will spread across the network at the fastest speed, so feedback on the information can be obtained in time. Moreover, the network makes it convenient to promote the globalization of research results so that applications for patents can be filed in many countries.

Fourth, MNCs are implementing systematic investment strategies. The foreign investment process of MNCs has gradually evolved from global marketization to globalization of production, technology and management. In addition to production projects, they are also investing in personnel training, technological development, marketing and other aspects to make full use of local raw materials, labor and intellectual resources and achieve their goal of reducing product costs. Meanwhile, the ability of products to adapt to the local markets is further strengthened.

Fifth, MNCs are seeking a sound R&D environment. R&D needs sound legal protection, technical facilities, conducive policies and socio-economic environment, as well as, more importantly, assistance from relevant technical fields. For MNCs, R&D is conducted in intelligence-intensive countries. Through extensive cooperation with local universities and research institutions, they outsource some of the R&D activities so that they can concentrate on R&D of their advantageous projects and products and further improve R&D efficiency.

Sixth, the R&D institutions serve as windows for technological information. Setting up R&D institutions in host countries, especially those with advanced technologies and those where their strong rivals are located, allows MNCs to better judge the development trend of technological frontiers, understand the dynamics of competitors and improve their efficiency in developing new technologies. Meanwhile, the institutions facilitate rational assessment of technological value as well as the introduction and purchase of technology. Such a purpose is especially evident when MNCs establish R&D institutions in the developing countries.

5.3.2. *Investment and revenue model*

First, the R&D internationalization investment of MNCs is determined as a rational behavior. Then, a set of mathematical models can be constructed to study the factors influencing the selection of location in the R&D internationalization of MNCs, such as consumption preferences, market capacity, the imitating ability of domestic enterprises and the protection of intellectual property rights, as well as the revenue under the corresponding conditions.

(1) Conditions and hypotheses: *First, the host country is a developing country and has a perfectly competitive market.* There is a local enterprise (VNC) and an MNC in the market. The technological level of the MNC (T_m) is ahead of that of the VNC (T_v), namely, $T_m > T_v$. When they produce the same kind of products, their qualities are Q_m and Q_v, respectively. Meanwhile, the quality of products is directly proportional to the technological level, i.e., $Q_m - Q_v = k(T_m - T_v)$, expressed as $\Delta Q = k\Delta T$, where k is the proportional coefficient between technology and quality.

Second, in the host country market, the price of product mainly depends on its quality and the preferences of consumers. In other words, the higher the product quality, the higher the price the consumers are willing to pay for it. U denotes the utility per unit of product purchased by the consumers, θ the preference coefficient of consumers for product quality and P the price of the product. Then, the utility per unit of product purchased by the consumers can be expressed as $U = \theta Q - P$, where θ is any value greater than 0.

Third, when the products are sold in the host country market, the enterprises are engaged in perfect competition and in equilibrium. Therefore, the average profit of the local enterprises is 0 and its production cost C_v is equal to the price P_v, i.e., the market price P^*. However, the price of MNCs' products P_m is decided by the consumers' preferences for product quality. When the consumers' preference coefficient for quality is $\theta^* = (P_m - P^*)/(Q_m - Q_v) = \Delta P/\Delta Q$, there is no difference in the consumers purchasing products from either side. However, when $\theta > \theta^*$, consumers buy products of MNCs. When $\theta < \theta^*$, consumers purchase the local enterprises' products.

Fourth, in the host country market, only MNCs invest in R&D. The investment amount is R and the investment efficiency is α. In this way, they maintain their leading role in technology. However, local enterprises do not invest in R&D but acquire relevant technologies by imitating the technology diffusion of MNCs. In theory, imitation also incurs cost, which is yet very low. It is assumed that such cost is ignored to simplify the model. At the same time, MNCs need protection measures for their R&D technologies to prevent host countries from imitating their technologies. The investment amount is I and the protection efficiency is β. They are dependent on the willingness of MNCs to protect the technologies and the imitation ability of host countries' local enterprises. Besides, the intensity of host countries in protecting their own intellectual property rights is γ. α, β and γ are all larger than 0 and smaller than 1, and $\alpha + \beta = 1$. A larger value indicates higher efficiency.

(2) **Derivation of correlation functions:** The first correlation function is the technology leadership function of MNCs According to the fourth

hypothesis and the Cobb–Douglas production function, the technological level of MNCs is $Tm = R^\alpha$. Host countries' intensity of legal protection and the effectiveness of MNCs' prevention of technology diffusion are $1 - [I^\beta(1 + \gamma)]^2$, i.e., $Tv = Tm\{1 - [I^\beta(1 + \gamma)]\}$. The technology leadership function of MNCs can be written as

$$\Delta T = Tm - Tv = Tm[I^\beta(1+\gamma)] = R^\alpha I^\beta(1+\gamma). \tag{1}$$

The second correlation function is the profit function of the MNCs in the host countries. The demand function for the MNCs' products in the host country market is expressed as

$$Dm = (\theta - \theta^*)\, D,$$

where D represents the total market capacity of the host countries. Therefore, the market profit of the MNCs in the host countries can be approximately expressed as

$$\pi_m = (P_m - P^*)D_m = \theta^* \Delta Q(\theta - \theta^*)D = \theta^*(\theta - \theta^*)D\Delta Q$$
$$= k\theta^*(\theta - \theta^*)D\Delta T.$$

It can be seen from the above analysis that $k\theta^*(\theta - \theta^*)D$ is directly related to the host country, which is reduced to Z here. The market profit of the MNCs in the host countries can be further expressed as $\pi_m = Z\Delta T$. That is, the relationship between market profit of the MNCs and their leading technology is linear.

With the cost of R&D considered, substitute the leading technology function of the MNCs (1) and then the net profit function of the MNCs in the host countries is

$$\pi_m^* = Z\Delta T - R - 1 = ZR^\alpha I^\beta(1+\gamma) - R - 1. \tag{2}$$

(3) Construction of model and equilibrium analysis: Generally speaking, the R&D investment activities conducted by the MNCs (C) include innovative investment (R) and protective investment (I) and $R + I = C$. Optimize the two so as to maximize the advantage of

---

[2]Li, A. F. *R&D Globalization of MNCs*. Beijing: People's Publishing House, 2004: 86.

leading technology ΔT. Then, combine the MNCs' market profits in the host countries with the R&D investment cost C to analyze the R&D decisions of the MNCs, i.e., under what conditions the market's net profit can be maximized.

$$\max_{R,I} \pi^{*}(R,I) = \pi_{M} - R - 1 = ZR^{\alpha} I^{\beta}(1+\gamma) - R - 1, \qquad (3)$$

where $R, I \geq 0$.

Solve Eq. (3). According to the Lagrange multiplier, the condition for maximizing π_{m}^{*} is

$$R = \frac{\alpha}{\alpha + \beta} C,$$

$$I = \frac{\beta}{\alpha + \beta} C.$$

Substitute the results into Eq. (2) and

$$\pi^{*} = Z \left(\frac{\alpha C}{\alpha + \beta} \right)^{\alpha} \left(\frac{\beta C}{\alpha + \beta} \right)^{\beta} (1+\gamma) - \frac{\alpha}{\alpha + \beta} C - \frac{\beta}{\alpha + \beta} C.$$

It can be reduced to

$$\pi^{*} = [Z\alpha^{\alpha}\beta^{\beta}(1+\gamma) - 1]C.$$

Substitute the parameters represented by Z and

$$\pi^{*} = [k\theta^{*}(\theta - \theta^{*})D\alpha^{\alpha}\beta^{\beta}(1+\gamma) - 1](R+I).$$

It can be seen from the above equation that the MNCs' establishment of R&D institutions and revenue is not only related to their technological conversion (k), investment and efficiency (R, I, α, β) but also directly linked to the market capacity (D) of the host countries, consumers' quality preferences (θ), local enterprises' level of quality and prices (θ^{*}), and intensity in protecting intellectual property rights (γ).

At present, the basic theoretical studies on the motivation behind R&D internationalization of the MNCs are mainly based on the results of the original theoretical research on the MNCs' FDI. Few have focused on the R&D activities. Since the behavior of R&D internationalization is more non-reciprocal and uncertain than the FDI of the MNCs, directly applying FDI theories to the R&D internationalization activities of the MNCs is not very pertinent. In addition, it is worth noting that these theoretical studies analyze mostly abstract theories based on the experience of developed countries. Therefore, these theories can better explain the behavior of R&D internationalization between developed countries. For developing countries with large technological gaps and relatively imperfect markets, the theories on R&D internationalization of the MNCs fail to offer convincing explanations. In response to this problem, the research in this book mainly focuses on the regions of China and examines the characteristics of the MNCs' R&D internationalization in China. Through the study of typical cases, it also sheds light on the contribution to and impact on the developing countries of such R&D internationalization.

Meanwhile, most of the academic research on the motivations behind R&D internationalization of the MNCs and their selection of location pays more attention to the benefits brought about by R&D internationalization, including the adaptability of target market technologies, access to R&D elements in host countries and security needs. It is difficult to analyze these studies under the same framework. For the MNCs, R&D internationalization is also an inherent requirement for effective and rational allocation of global resources under the goal of maximizing global profits. Therefore, how to incorporate the agreement of MNCs' R&D internationalization into the decision for maximizing global profit has always been a tough problem. At present, the academic research on it is also insufficient. More importantly, the basic starting point of existing theoretical research is that the behavior of R&D internationalization of the MNCs is generally made between countries (or enterprises) with a similar technological level. However, in real life, a large number of MNCs have extended the scope of R&D internationalization to new industrialized countries and developing countries. It can be seen that it is difficult for the existing theories to explain the vertical cooperation behavior in the R&D internationalization of the MNCs. This book conducts an in-depth study of the problem from an empirical perspective. By constructing models for the efficiency in R&D internationalization and its effect on the level, we analyze the mechanisms by which the MNCs conduct R&D internationalization in developing countries and how they decide the level of R&D internationalization.

Studies on the spillover effects of R&D internationalization of the MNCs are mainly normative research into the spillover channels. Most of the scholars believe that the spillover channels include the human capital linkage effect, the technological R&D demonstration effect, the product market competition effect and the R&D location agglomeration effect. However, few have discussed the internal links between different spillover channels. The research in this book mainly specifies the spillover effects of the MNCs' R&D internationalization on the indigenous innovation of host countries from two aspects, namely, the promoting effect and the inhibitory effect.

Furthermore, the current academic research on indigenous innovation is still in its infancy. There is hardly any study on MNCs' R&D internationalization as an important channel to enhance the capacity of indigenous innovation. This aspect has been ignored by the academic community. Therefore, the research in this book will fill the gaps in domestic and foreign research in this aspect. Through theoretical analysis, we clarify the positive and negative effects of the MNCs' R&D internationalization behavior on indigenous innovation and build an indigenous innovation output model to analyze how such behavior contributes to the indigenous innovation of the host countries.

Chapter 2

Research on the Level and Efficiency of R&D Internationalization of MNCs

According to the definition in the Frascati Manual (OECD, 1964), R&D "comprise creative and systematic work undertaken in order to increase the stock of knowledge — including knowledge of humankind, culture and society — and to devise new applications of available knowledge".[1] Be it the neoclassical economic growth theory or the endogenous economic growth theory, it is believed that R&D activities are crucial to technological progress and improvement in productivity in the long run as well as form an important source of impetus for economic growth.

Due to the huge technological spillover effect of R&D activities, for a long time, the MNCs have only shifted the low value-added parts of the value chain outwards to maintain their core competitive advantages and have concentrated their R&D institutions in the home countries. However, with the development of market integration and the rise of knowledge economy, the MNCs have gradually changed this tradition in order to enhance their international competitiveness and began to set up R&D institutions overseas (Xi and Ge, 2000), which has fostered the new trend — R&D internationalization — under the new system of economic globalization.

---

[1]Currently, China adopts the UNESCO definition. R&D activities refer to the systematic and creative activities conducted in order to increase the stock of knowledge, including knowledge of humankind, culture and society, and to devise new applications of such knowledge. They comprise basic research, applied research and experimental development (also known as technological development).

At the same time, with the development of market economy and opening up, China has become an important host country that attracts cross-border R&D investment with its huge market size, fast-growing economy and abundant resources. By the end of 2006, there had been 1,779 large and medium-sized industrial enterprises that belong to the three types of foreign-funded enterprises (Sino-foreign joint ventures, Sino-foreign cooperatives and wholly foreign-funded enterprises) and have set up development institutions in China, with a total of 2,223 technological institutions in the country. In 2006, the internal expenditures on technological activities of the three types of enterprises in China added up to ¥84.818 billion. The total amount of technological R&D funds raised was ¥90.276 billion.[2] The increase in R&D expenditure of the three types of foreign-funded enterprises points to the close attention paid by the enterprises to the R&D behavior. From 1998 to 2006, the annual growth in the amount of technological development funds raised by large and medium-sized industrial enterprises that belong to the three types of foreign-funded enterprises in China was 33.32% and that in the internal expenditure of technological development funds increased by 35.13%. Meanwhile, we find that, behind the huge investment in R&D expenditure, the R&D results also show a trend of significant growth. From 1998 to 2006, the annual growth in gross industrial output value of the large and medium-sized industrial enterprises that belong to the three types of foreign-funded enterprises in China was 34.10% and that in the number of granted patents was 53.02%, which is apparently higher than the growth in the funds invested in R&D internationalization over the same period. Therefore, it is necessary to analyze the current development of the level and efficiency of R&D internationalization in China and find out the relationship between input and output in R&D internationalization. In 2018, China's R&D expenditures reached ¥1,965.7 billion, up by 11.6% compared with those in 2017 and accounting for 2.18% of the GDP. They have promoted the development of new economy in China. In 2018, the value-add of China's strategic emerging industries increased by 8.9% compared with that in 2017. The added values of high-tech manufacturing industry and equipment manufacturing industry rose by 11.7% and 8.1%, respectively.

---

[2] *Data source: China Statistical Yearbook on Science and Technology* from 2004 and 2007, covering data from 2003 to 2006.

On the basis of these facts, we believe how to measure the level and efficiency of R&D internationalization and examine the mechanism by which different locational factors impact the level of R&D internationalization, affecting the efficiency in R&D internationalization, is a major question that needs to be answered, so that China can deliver the goals of rationally absorbing and utilizing overseas R&D resources, pursuing an R&D internationalization model with high efficiency and fast growth, improving the level of R&D internationalization in each region, and ultimately promoting technological progress, productivity and economic growth.

1. Research on the Level and Efficiency of R&D Internationalization

In terms of the relationship between the level and efficiency of production activities, the academic community mainly focuses on the overall study of the level of production and tends to overlook research on efficiency. Although there are only a few studies on the production activities of R&D internationalization, their basic ideas and research methods are consistent. The sources of economic growth are divided into four parts: capital accumulation, employment growth, human capital growth and total factor productivity (TFP) growth. TFP growth promotes production activities through technological progress and technological efficiency. Existing studies on the TFP of economic aggregate mainly used time-series data on economic aggregate and tend not to distinguish between technological progress and technological efficiency. Besides, because of the small size of time-series data used, it is difficult to choose a complex functional form to estimate the production function and divide productivity. Some studies divide productivity on the basis of interprovincial panel data (Wu, 2000). Among these studies, Wu uses a stochastic frontier panel data production model while Lin and Liu adopt the Malmquist index to calculate the rate of technological progress.

In terms of the performance evaluation of R&D internationalization, foreign research dates back to the 1980s with four main theories: classical model evaluation theory (Jackson, 1983; Schmitt and Freeland, 1992), organizational decision-making theory (Sounder and Mandakovic, 1986), strategic performance theory (Moon and Bates, 1993), and option pricing theory (Newton and Pearson, 1994). These models mainly measure R&D activities from the perspective of organization or project without macroscopic measurement of R&D internationalization from a national or regional perspective. In particular, they hardly touch upon the concept of internationalization.[3] They focus on the quantitative output of R&D activities and attach too much attention to the direct output and short-term results. Many potential factors that are beneficial to long-term development and take a certain period of time to play a part are often ignored. As for the quantitative evaluation, many of these evaluations at home and abroad rely on individual subjective experience to give weight to each indicator without scientific and systematic methods for support. Thus, it is difficult to ensure the validity of evaluation results.

As regards the factors influencing R&D internationalization, early research finds that the market size of host countries is the most important factor affecting overseas R&D investment. Mansfield *et al.* (1979) discover that when the sales of overseas subsidiaries take up a larger proportion of the total sales of the MNCs, the MNCs spend more money on the R&D by the subsidiaries. Hewitt (1980) finds that the more the export activities to the home countries the overseas foreign subsidiaries are engaged in, the more likely the MNCs are to conduct R&D activities there. Zejan (1990) believes that the market size and per capita income of the host countries have a positive effect on the R&D investment of the MNCs. A growing number of studies have found that access to technical resources exerts an increasingly apparent impact on the overseas R&D investment. Miller's (1993) study shows that the main motivation for overseas R&D is to ensure that technologies remain advanced. The study by Robert Pearce (1999) reveals that market competition and product innovation are the main factors affecting overseas R&D investment. Dalton and Serapio (1999) argue that the availability of knowledge and talent in a country will have an impact on its attraction to overseas R&D investment. Infrastructure and policy environments are also important factors. Kumar (1996) finds that overseas R&D of the American MNCs has a close connection with the host countries' policy environment, especially the protection of intellectual property rights. Lin and Chai (1998) believe that cross-investment in high-tech fields tends to be made in host

---

[3] The work has drawn great attention from some countries. OECD has begun the study of the indicator system for the measurement of R&D internationalization, and this is still in the tentative early stages.

countries with less strict research policies, better service facilities, and huge market potential. Liu (2003) believes that the MNCs often set up overseas R&D institutions in areas that enjoy strong R&D strength, feature industrial advantages or gather a lot of manufacturers. Besides, urban infrastructure construction and regional preferential policies can greatly attract overseas R&D investment.

2. Measurement of the Level and Efficiency of R&D Internationalization

2.1. *Measurement of the overall level of R&D internationalization*

In this book, the overall level of regional R&D internationalization is assessed through the principal component analysis method. The main purpose of this method is to interpret most of the variable information in the original data with less variables and to convert many of the highly relevant variables in our hands into variables that are independent of or unrelated to each other. Usually, several new variables whose number is smaller than that of the original variables and which can account for the variation in most of the data, i.e., the principal components, are selected as the comprehensive indicators for interpreting the data. Thus, the principal component analysis method is actually a dimensionality reduction method.

In the empirical studies, in order to analyze problems comprehensively and systematically, we must consider a number of influencing factors. Because each variable reflects certain information of the problem under study to a varying degree and the indicators have a certain correlation with each other, the information reflected by the statistics obtained may overlap to some extent. While using statistical methods to study multivariate problems, too many variables will increase the amount of calculation as well as the complexity of the analysis. The principal component analysis method allows us to get more information with fewer variables.

The main principle of the principal component analysis method is to try to replace the original indicators with a new set of mutually independent indicators by recombining the original indicators related to each other to a certain degree (such as m indicators). Mathematically, the usual processing method is to linearly combine the *m* original indicators into new comprehensive indicators. The specific process is as follows:

(1) Collect *m* evaluation indicators in *n* regions to form an $n \times m$ matrix X_{ij} ($i = 1, 2, \ldots, n; j = 1, 2, \ldots, m$). Since the units of indicators are different, the effect of dimension should be eliminated. Standardize the raw data:

$$ZX_{ij} = (X_{ij} - \overline{X_j})/\delta_j,$$

where $X_j = \frac{1}{n}\sum_{i=1}^{n} X_{ij}$, $\delta_i = \sqrt{\frac{1}{n}(X_{ij} - \overline{X_j})^2}$.

(2) Calculate the correlation coefficient matrix R_{jk} of the dimensionless data:

$$(R_{jk})_{m \times n} = \frac{1}{n}\sum_{i=1}^{n} ZX_{ij} \cdot ZX_{jk} \quad (j,k = 1,2,\ldots,m).$$

(3) Starting from R_{jk}, figure out the eigenvalue λ_j and the corresponding eigenvector I_{jk}. The percentage of each eigenvalue to the total is the variance contribution rate of each principal factor P_j:

$$P_j = \frac{\lambda_j}{\sum_{j=1}^{m} \lambda_j}.$$

(4) Determine the number of principal components whose cumulative variance contribution rate is greater than 85%. Based on the correlation between the *k*th principal component F_k and the original component vector ZX_j, calculate the factor loading matrix R_{jk} of the *k*th principal factor:

$$R_{jk} = \sqrt{\lambda_k} \cdot I_{jk} \quad (j = 1,2, \ldots, m; k = 1,2, \ldots, p; p \leq m).$$

On the basis of the above, we can calculate the score of the principal components F_{ik} and the overall scores Z_i of each region's investment environment:

$$F_{ik} = P_{kj} / \sqrt{\lambda} k \cdot ZX_{ij};$$

$$Z_i = \sum_{k=1}^{p} \sum_{j=1}^{m} P_k \cdot F_{ik}.$$

This method avoids the subjectivity and arbitrariness in determining weight when methods such as overall scoring are used and allows the evaluation results to be in line with the actual situation. Meanwhile, the principal components are represented as a linear combination of the original variables. If the final comprehensive indicators include all components, then the accurate results can be obtained and 100% of the variation information provided by the original variables can be retained. Even if we discard a few components, we can guarantee that more than 85% of the variation information is reflected in the overall score, so that the evaluation results are true and reliable. This is a widely used method in practice. Since the first principal component (factor) contains the largest amount of information in all principal components, many scholars often use it to compare the differences between different entities while studying the problem of comprehensive evaluation. To sum up, the method mainly has two strengths: (1) objectivity in determining weight; (2) true and reliable results of evaluation.

2.2. Construction of indicator system for level of R&D internationalization and measurement

Based on the viewpoints given in the previous section, for a country or region, the general level of R&D internationalization is the sum of the level of a foreign entity (enterprise, government, university, international organization, etc.) in conducting R&D activities and creating R&D output and that of local entities of a region that have gone international in conducting R&D activities and creating R&D output abroad. In a narrow sense, the level of R&D internationalization refers to the level of the foreign MNCs making comprehensive use of creative resources of two or more countries in a region, such as capital, institutions, knowledge input, and technological talent, to carry out R&D activities and create R&D output. Given China's current technological and economic development and since Chinese companies have been engaged in overseas R&D investment only since 2000, with a small scale and quantity and the

MNCs are always the most dynamic backbone in the process of R&D internationalization, this book focuses on the level of R&D internationalization of the MNCs in China, namely, the level of international R&D in the narrow sense.

Since no authoritative organization has released indicators to measure the level of R&D internationalization, this book draws on Huang's (2006) design of indicators and makes corresponding improvement and correction. From the macro perspective of countries and regions, this book divides R&D internationalization into four primary indicators, namely, capital internationalization, institution internationalization, activity internationalization and output internationalization, as well as 23 secondary indicators to specifically measure the level of R&D internationalization in different regions of China (see Table 1). Since the relevant statistics in China do not include all the foreign-funded enterprises but only the large and medium-sized industrial enterprises (three types of foreign-funded enterprises), measurement can only be made according to this caliber. Hence, there is a certain degree of underestimation in statistical analysis.

We use the SPSS software to measure the level of R&D internationalization of 30 provinces/municipalities/autonomous regions in China based on principal component analysis. Based on the results of measurement, we calculate the average level of R&D internationalization of 30 regions in China from 2003 to 2008, as well as that of the eastern, central and western regions, as shown in Table 2 and Figure 1.[4]

As indicated in Figure 1, the national level of R&D internationalization has increased year by year. The level of R&D internationalization is higher in the eastern region and generally lower in the central and western regions. The level of R&D internationalization has remained high over the years in Guangdong, Shanghai and Jiangsu, followed by Zhejiang, Fujian and Tianjin. These provinces with a higher level of R&D internationalization are located in the eastern coastal areas and are consistent with other regions in the attraction of overseas R&D investment and the level of economic development. However, Beijing, despite being an important

---

[4]The eastern region includes 11 provinces and municipalities, including Beijing, Tianjin, Hebei, Liaoning, Shanghai, Jiangsu, Zhejiang, Fujian, Shandong, Guangdong and Hainan. The central region includes eight provinces, including Shanxi, Jilin, Heilongjiang, Anhui, Jiangxi, Henan, Hubei, and Hunan. The western region includes 12 provinces, municipalities and autonomous regions, including Chongqing, Sichuan, Guizhou, Yunnan, Tibet, Shaanxi, Gansu, Qinghai, Ningxia, Xinjiang, Guangxi and Inner Mongolia.

Table 1. Indicator System for Level of R&D Iinternationalization

Primary indicator	No.	Secondary indicator	Unit
Capital internationalization	1	Amount of technological development fund raised by large and medium-sized industrial enterprises (three types of foreign-funded enterprises)	¥10,000
	2	Ratio to national fund for technological activities	%
	3	Amount of internal expenditure on technological development of large and medium-sized industrial enterprises (three types of foreign-funded enterprises)	¥10,000
	4	Ratio to FDI	%
	5	Expenditure on technological transformation of large and medium-sized industrial enterprises (three types of foreign-funded enterprises)	¥10,000
	6	Expenditure on technological introduction of large and medium-sized industrial enterprises (three types of foreign-funded enterprises)	¥10,000
	7	Expenditure on assimilation and absorption of large and medium-sized industrial enterprises (three types of foreign-funded enterprises)	¥10,000
	8	Expenditure on purchase of domestic technology of large and medium-sized industrial enterprises (three types of foreign-funded enterprises)	¥10,000
Institution internationalization	9	Number of staff in technological development institutions of large and medium-sized industrial enterprises (three types of foreign-funded enterprises)	Person
	10	Number of technological development institutions of large and medium-sized industrial enterprises (three types of foreign-funded enterprises)	—
	11	Number of large and medium-sized industrial enterprises (three types of foreign-funded enterprises) that have technological development institutions	—
	12	Ratio to large and medium-sized industrial enterprises (three types of foreign-funded enterprises)	%

(Continued)

Table 1.   (*Continued*)

Primary indicator	No.	Secondary indicator	Unit
Activity internationalization	13	Number of engineers and technicians of large and medium-sized industrial enterprises (three types of foreign-funded enterprises)	Person
	14	Ratio to employees of large and medium-sized industrial enterprises (three types of foreign-funded enterprises) at year end	%
	15	Number of technological developers of large and medium-sized industrial enterprises (three types of foreign-funded enterprises)	Person
	16	Number of patent applications of large and medium-sized industrial enterprises (three types of foreign-funded enterprises)	—
	17	Number of technological development projects of large and medium-sized industrial enterprises (three types of foreign-funded enterprises)	—
	18	Number of new product development projects of large and medium-sized industrial enterprises (three types of foreign-funded enterprises)	—
Output internationalization	19	Sales revenue from new products of large and medium-sized industrial enterprises (three types of foreign-funded enterprises)	¥10,000
	20	Ratio to product sales revenue of large and medium-sized industrial enterprises (three types of foreign-funded enterprises)	%
	21	Gross industrial output value of new products of large and medium-sized industrial enterprises (three types of foreign-funded enterprises)	¥10,000
	22	Ratio to the gross industrial output value of large and medium-sized industrial enterprises (three types of foreign-funded enterprises)	%
	23	Number of patents granted of large and medium-sized industrial enterprises (three types of foreign-funded enterprises)	—

Table 2.   Comprehensive Indicators for Level of R&D Internationalization

Area	2003	2004	2005	2006	2007	2008
Beijing	17.83	25.61	20.52	20.20	21.69	24.91
Tianjin	29.07	32.70	47.23	41.23	47.31	61.25
Hebei	8.19	9.00	11.50	15.68	19.55	29.19
Shanxi	8.52	8.55	7.88	4.74	3.99	2.73
Inner Mongolia	15.99	12.69	12.53	11.48	10.32	8.24
Liaoning	9.15	10.30	10.51	11.38	12.25	14.12
Jilin	3.08	0.12	16.56	24.88	26.70	24.14
Heilongjiang	8.54	12.07	11.02	13.53	16.03	21.95
Shanghai	100.00	74.42	80.13	80.58	75.92	67.14
Jiangsu	50.69	100.00	65.59	68.67	79.00	78.40
Zhejiang	26.12	64.14	45.34	48.40	68.24	60.80
Anhui	18.78	14.34	26.71	26.45	31.89	44.99
Fujian	41.74	37.22	45.63	52.57	57.30	67.60
Jiangxi	14.85	3.40	12.47	15.37	26.28	63.58
Shandong	17.79	19.68	17.76	25.16	28.73	36.87
Henan	21.15	18.68	13.97	18.47	18.18	17.61
Hubei	10.51	11.81	23.05	13.59	16.60	23.96
Hunan	4.11	6.55	10.47	7.57	9.88	15.91
Guangdong	71.87	75.18	100.00	72.67	75.16	80.32
Guangxi	12.06	8.83	12.17	6.28	5.50	4.13
Hainan	5.09	2.34	9.23	5.29	8.78	20.35
Chongqing	14.30	12.10	13.02	12.78	12.37	11.58
Sichuan	11.29	8.56	14.72	9.14	9.44	10.06
Guizhou	13.46	9.57	11.10	22.51	29.25	30.80
Yunnan	9.37	5.79	10.29	9.53	10.55	12.81
Shaanxi	5.80	6.66	11.94	11.35	14.72	23.48
Gansu	7.49	6.55	8.18	7.78	7.97	8.37
Qinghai	0.03	0.02	7.28	0.01	0.76	2.10
Ningxia	6.68	18.71	0.07	16.90	20.00	22.30
Xinjiang	3.39	2.08	8.02	0.75	1.14	2.33

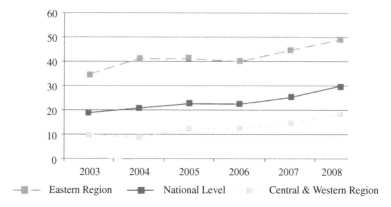

Figure 1.   Comparison Between the Eastern, Central and Western Regions in Terms of R&D Internationalization.

location for overseas R&D investment and the center of education and culture in China, has a lower level of R&D internationalization.

2.3. *Measurement of the efficiency in R&D internationalization based on DEA*

Data envelopment analysis (DEA) is a useful decision-making method for evaluating the relative effectiveness of similar departments or units (decision-making units, DMUs), including the CCR model (Charnes *et al.*, 1978) and the BCC model (Banker *et al.*, 1984). The bases for the evaluation include the necessary amount consumed in the activity, i.e., the so-called input indicator, and the amount that indicates the effectiveness of the activity, i.e., the so-called output indicator. In nature, the efficiency is measured by using a mathematical programming model to estimate the frontier of effective production and then comparing each DMU with such frontier based on a set of observed value of input and output. For DMUs on the frontier, DEA regards their input–output combinations as the most efficient and sets their efficiency indicator as 1. DMUs that are not on the frontier are considered inefficient and relative efficiency indicators (greater than 0 and smaller than 1) are given with the effective point of efficiency frontier as the benchmark.

Suppose that there are two kinds of R&D inputs and one kind of R&D output (L, K, Y). Take the average of labor and obtain (k, y), where $k = K/L$ and $y = Y/L$. Suppose that there are two periods: the base

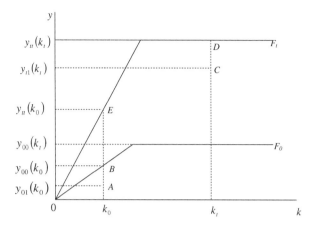

Figure 2. Decomposition of Economic Growth.

period and the t period. In the base period, the current factors can be used to obtain the maximum output level under the established R&D conditions. Such a trajectory is the boundary of the R&D frontier of the base period, which is denoted as F_0. However, since the adequacy of the role of technology and that of the factor input may not be matched, the actual organizational and productive capabilities are often lower than the boundary of production frontier, as shown by point A in Figure 2. The difference between the two can be considered as the difference in technological efficiency. In other words, under the same R&D factor input k_0, the technologies cannot be fully utilized and the R&D output cannot reach the boundary of the frontier. The same thing happens on Ft in the t period. Denote the technological efficiency above as e. As shown in Figure 2, when the R&D factor input in the base period is k_0, y_{01} and y_{00} are the corresponding actual R&D output and maximum output, respectively. Then, we have $e_0 = \frac{y_{01}(k_0)}{y_{00}(k_0)}$. Similarly, we have $e_t = \frac{y_{t1}(k_t)}{y_{tt}(k_t)}$ in the t period. Then,

$$\frac{y_{t1}(k_0)}{y_{01}(k_0)} = \frac{e_t \cdot y_{tt}(k_t)}{e_0 \cdot y_{00}(k_0)}. \tag{1}$$

If we use the technology in base period to measure the deepening of R&D capital, then the above formula can be written as follows:

$$\frac{y_{t1}(k_0)}{y_{01}(k_0)} = \frac{e_t}{e_0} \cdot \frac{y_{tt}(k_t)}{y_{0t}(k_t)} \cdot \frac{y_{0t}(k_t)}{y_{00}(k_0)}. \tag{2}$$

In this way, R&D output growth is broken down into three parts: improvement in technological efficiency e_t/e_0, technological progress $y_{tt}(k_t)/y_{0t}(k_t)$ and capital deepening $y_{0t}(k_t)/y_{00}(k_0)$. If the technological progress is measured by the capital of the base period and the capital deepening by the technology of the t period, then Eq. (2) can be written as follows:

$$\frac{y_{t1}(k_0)}{y_{01}(k_0)} = \frac{e_t}{e_0} \cdot \frac{y_{tt}(k_t)}{y_{t0}(k_0)} \cdot \frac{y_{t0}(k_0)}{y_{00}(k_0)}. \tag{3}$$

Both Eqs. (2) and (3) can decompose the growth of R&D output. Unless technology is Hicks-neutral, the results of the two ways of decomposition are not equal. Hence, we will follow Caves *et al.* (1982) and use the geometric mean for capital deepening and technological progress:

$$\frac{y_{rc}}{y_{rb}} = \frac{e_c}{e_b} \cdot \left(\frac{y_{tt}(k_t)}{y_{0t}(k_t)} \cdot \frac{y_{0t}(k_t)}{y_{00}(k_0)} \right)^{\frac{1}{2}} \left(\frac{y_{tt}(k_t)}{y_{t0}(k_0)} \cdot \frac{y_{t0}(k_0)}{y_{00}(k_0)} \right)^{\frac{1}{2}}. \tag{4}$$

For specific calculation, the Malmquist index can be used to calculate the TFP change based on the R&D output and input of each region and at the same time economic growth can be decomposed into two parts, namely, improvement in efficiency and technological progress. On this basis, the capital deepening is calculated based on the growth rate of R&D output.

In this book, DEAP2.1[5] software is applied to measure the efficiency of R&D internationalization. During the calculation, a total of 30 DMUs are selected. They are the provinces/autonomous regions/municipalities of China, except Tibet. There are three input variables: the amount of internal expenditure on technological development, the expenditure on other technological activities and the technological developers of large and medium-sized industrial enterprises (three types of foreign-funded enterprises). There are three output variables: the gross industrial output value of new products, the number of technological development projects, and

---

[5]Tim Coelli, University of New England. Since there is no DEA software in China and DEAP2.1 is commonly used to calculate efficiency, we use it for measuring the efficiency of R&D internationalization in this book.

the number of patents granted of large and medium-sized industrial enterprises (three types of foreign-funded enterprises). The number of data periods selected for input is 5, i.e., from 2002 to 2008. During the calculation of the Malmquist index, the production technologies in the previous period is used as a basis of reference, so the data for the first period are lost. Table 3 shows the measurement results of efficiency of R&D internationalization from 2007 to 2008.

Based on the results for the efficiency of R&D internationalization from 2007 to 2008, we find that regions with a higher level of R&D internationalization have a higher regional technological efficiency indicator Effch and lower technological progress indicator Tech. On the contrary, regions with a lower level of R&D internationalization have a lower regional technological efficiency indicator Effch and a higher technological progress indicator Tech. These conclusions are drawn from the surface data characteristics, and which offers an accurate judgment.

3. Mechanism Behind the Influence of Efficiency on the Level of R&D Internationalization

3.1. *Construction of models*

R&D internationalization is a creative production process where the knowledge of human beings, culture and society is used to make new inventions and continuously enrich the stock of knowledge. The fundamental factor affecting the level of production is the efficiency of the production activity. Therefore, we draw on the output model to construct the following model for the relationship between the level and efficiency of R&D internationalization:

$$I = \alpha \, \text{Tech}^{\beta_1} \text{Pech}^{\beta_2} \text{Sech}^{\beta_3} \mu, \tag{5}$$

where I is the comprehensive indicator of R&D internationalization and is derived from the actual measurement above. Tech, Pech, and Sech are the efficiency in technological progress, pure technological efficiency and scale efficiency of R&D internationalization, respectively. They are variables without units.

We use panel data analysis method. In order to compare the empirical results, we take the logarithm of both sides of Eq. (5) and obtain the

Table 3.   Efficiency of R&D Internationalization Between 2007 and 2008

DMU	Effch	Tech	Pech	Sech	TFPch
Beijing	1.000	0.843	1.000	1.000	0.843
Tianjin	2.791	0.384	1.000	2.791	1.072
Hebei	3.330	0.264	1.659	2.007	0.880
Shanxi	3.775	0.239	2.967	1.272	0.900
Inner Mongolia	6.268	0.290	5.515	1.137	1.817
Liaoning	4.951	0.270	1.984	2.496	1.338
Jilin	1.000	0.078	1.000	1.000	0.078
Heilongjiang	4.896	0.248	3.043	1.609	1.212
Shanghai	2.814	0.326	1.000	2.814	0.916
Jiangsu	2.466	0.402	1.057	2.334	0.993
Zhejiang	4.142	0.347	1.570	2.639	1.438
Anhui	2.476	0.308	1.459	1.697	0.762
Fujian	3.740	0.396	1.363	2.744	1.480
Jiangxi	4.355	0.303	2.704	1.611	1.319
Shandong	2.989	0.376	1.169	2.556	1.125
Henan	3.931	0.291	2.594	1.516	1.143
Hubei	2.537	0.303	1.265	2.006	0.768
Hunan	1.000	0.255	1.000	1.000	0.255
Guangdong	2.259	0.472	1.000	2.259	1.065
Guangxi	2.618	0.281	1.443	1.815	0.735
Hainan	0.812	0.258	0.861	0.944	0.209
Chongqing	3.662	0.336	1.346	2.721	1.229
Sichuan	2.782	0.292	2.093	1.329	0.813
Guizhou	1.000	0.544	1.000	1.000	0.544
Yunnan	3.967	0.309	2.911	1.363	1.225
Shaanxi	1.380	0.346	1.315	1.049	0.478
Gansu	2.702	0.355	2.657	1.017	0.961
Qinghai	1.153	0.359	0.414	2.789	0.414
Ningxia	3.840	0.303	3.570	1.076	1.165
Xinjiang	7.841	0.314	8.940	0.877	2.461

logarithmic model (6). In this way, the coefficients before the explanatory variables represent concepts of elasticity.

$$\ln I_{it} = \alpha + \beta_1 \ln \text{L} n \text{Tech}_{it} + \beta_2 \ln \text{Pech}_{it} + \beta_3 \ln \text{Sech}_{it} + \mu. \qquad (6)$$

3.2. *Empirical test of the influence of efficiency on the level of R&D internationalization*

We first examine 30 interprovincial samples from China and analyze the overall effect of efficiency on the level of R&D internationalization. Since there are huge differences in the development indicators between eastern and central/western regions in China, the impact of efficiency on the level of R&D internationalization of the two regions should also be different. To test such differences, all the samples are divided into two groups based on eastern and central/western regions for the analysis of efficiency's influence on the level of R&D internationalization. After F test and Hausman test, the three equations are all analyzed with individual fixed effects models. The results are shown in Table 4.

At the national level, technological progress, pure technological efficiency and scale efficiency significantly promote the level of R&D internationalization, which is in line with the theoretical viewpoint. Among them, the positive effect of technological progress and scale efficiency is significant at the level of 5%, while the pure technological efficiency is significant at the level of 10%. As regards the coefficient of explanatory variables, for every 1% of increase in the technological progress rate, the level of R&D internationalization goes up by 0.179%. For every 1% of increase in pure technological efficiency, the level of R&D internationalization rises by 0.159%. For 1% of increase in scale efficiency, the level of R&D internationalization rises by 0.076%. It can be seen that technological progress and pure technological efficiency are more significant than scale efficiency in promoting the level of R&D internationalization. In other words, improvement in the overall level of R&D internationalization in China is mainly dependent on technological progress and pure technological efficiency. Scale efficiency makes smaller contributions. Based on the intercepts in the regression model, the situation varies across regions. Guangdong, Shanghai, Jiangsu, Zhejiang, Fujian, and Tianjin have more obvious fixed effects and they are located in those parts of the

Table 4.   Regression Results of Individual Fixed Effects Models (2003–2008)

	Overall regression results		Regression results of eastern regions		Regression results of central/western regions	
	Coefficient	T-test value		Coefficient	T-test value		Coefficient	T-test value
ln Tech	0.719	3.629*	ln Tech	0.220	1.312	ln Tech	0.213	3.609*
ln Pech	0.076	1.526**	ln Pech	0.397	2.904*	ln Pech	0.063	1.101
ln Sech	0.159	2.720	ln Sech	0.289	2.578*	ln Sech	0.057	0.517
C	2.501	23.489	C	4.001	28.310*	C	2.251	37.080*
Adjusted R^2	0.978	—	Adjusted R^2	0.980	—	Adjusted R^2	0.921	—

Note: * and ** refer to significance at the level of 5% and 10% (applicable also to other tables).

eastern region that are more economically developed. This is consistent with the previous assumption and provides a basis for regional tests.

According to the regional regression results, the effect of efficiency on the level of R&D internationalization is mainly reflected in pure technological efficiency and scale efficiency, i.e., technological efficiency, in the eastern region it's mainly derived from technological progress in the central/western regions. As the results of the eastern region indicate, technological efficiency plays a significant role in promoting the level of R&D internationalization. The level of R&D internationalization increases by 0.397% for every 1% of increase in pure technological efficiency and by 0.289% for every 1% of increase in scale efficiency. Both are significantly larger than the results from analysis on data at national level. Technological progress has no obvious promotional effect on the level of R&D internationalization in the eastern region. In the central/western regions, technological progress has contributed the most to the improvement in the level of R&D internationalization, while pure technological efficiency and scale efficiency have no significant effect. The level of R&D internationalization rises by 0.213% for every 1% of increase in technological progress in the central/western regions, which is significantly greater than the effect of technological progress at the national level. The intercept coefficients of the two models show that the fixed effect in the eastern region is greater than that in the central/western regions, which is consistent with the results obtained by the national model.

The regression results of the three models taken together, technological progress, pure technological efficiency and scale efficiency, have significantly promoted the level of R&D internationalization in China. Among these, the improvement in the level of R&D internationalization in the eastern region is mainly derived from technological progress while that in the central/western regions is mainly affected by technological efficiency, namely, pure technological efficiency and scale efficiency.

3.3. *Factors influencing the efficiency's effect on the level of R&D internationalization*

According to the previous studies, the regional market size, scale of foreign investment, degree of protection for intellectual property rights, R&D resources and infrastructure construction are external factors that have an important impact on R&D internationalization. Therefore, we examine the factors influencing how efficiency affects the level of R&D internationalization in these five aspects. The indicators are detailed in the following sections.

3.3.1. Market size

We use the annual average of GDP of the provinces from 2003 to 2008 (unit: ¥100 million) to measure their market size. The 30 provinces are divided into two groups based on the annual average of GDP. The first group consists of 15 provinces, including Guangdong, Shandong, Jiangsu, Zhejiang, Henan, Hebei, Shanghai, Liaoning, Sichuan, Hubei, Fujian, Hunan, Heilongjiang, Anhui and Beijing. The second group comprises 15 provinces, including Jiangxi, Guangxi, Shanxi, Jilin, Tianjin, Shaanxi, Yunnan, Inner Mongolia, Chongqing, Xinjiang, Guizhou, Gansu, Hainan, Ningxia and Qinghai. The annual average of GDP of the former group is higher than that of the latter.

3.3.2. Scale of foreign investment

We use the annual average of FDI stock of the provinces from 2003 to 2008 (unit: $10,000) to measure their scale of foreign investment. With the actually utilized FDI flow indicator of the regions, we adopt "perpetual inventory method" to measure the FDI stock of the regions. For the stock in base period, the investment flow of the base year is divided by 10% (Zhang, 2004; Young, 2000) for estimation. Considering the availability of data, we draw on the above method, with 1985 as the base year,[6] calculate the stock of 1985 and use it as the base stock. We also assume a depreciation rate of 9.6% and figure out the FDI stock from 2003 to 2008 based on the "perpetual inventory method". We divide the 30 provinces into two groups based on the annual average of FDI. The first group consists of 15 provinces, including Guangdong, Jiangsu, Shandong, Fujian, Shanghai, Zhejiang, Liaoning, Beijing, Tianjin, Hubei, Hebei, Hunan, Jiangxi, Heilongjiang and Hainan. The second group comprises 15 provinces, including Henan, Guangxi, Sichuan, Shaanxi, Anhui, Chongqing, Jilin, Inner Mongolia, Shanxi, Yunnan, Guizhou, Gansu, Qinghai, Xinjiang and Ningxia. The annual average of FDI of the former group is higher than that of the latter.

---

[6]Data on Beijing, Inner Mongolia, Hubei, Qinghai and Sichuan are lost. Hence, the year with data is used as the base year for calculation using the same method.

3.3.3. *Degree of protection for intellectual property rights*

We use the annual average of patents owned by every 10,000 people of the provinces from 2003 to 2008 (unit: per 10,000 people) to measure their degree of protection for intellectual property rights. We divide the 30 provinces into two groups based on the annual average of patents owned by every 10,000 people. The first group consists of 15 provinces, including Shanghai, Beijing, Guangdong, Zhejiang, Tianjin, Jiangsu, Fujian, Liaoning, Chongqing, Shandong, Heilongjiang, Jilin, Hubei, Hunan and Sichuan. The second group comprises 15 provinces, including Hebei, Shaanxi, Ningxia, Xinjiang, Henan, Shanxi, Inner Mongolia, Yunnan, Hainan, Jiangxi, Anhui, Guangxi, Guizhou, Gansu and Qinghai. The annual average of patents owned by every 10,000 people of the former group is higher than that of the latter.

3.3.4. *R&D resources*

We use the annual average of the R&D funds of the society at large of the provinces from 2003 to 2008 (unit: ¥10,000) to measure their abundance of R&D resources. We divide the 30 provinces into two groups based on the annual average of the R&D funds of the society at large. The first group consists of 15 provinces, including Beijing, Jiangsu, Guangdong, Shanghai, Shandong, Zhejiang, Sichuan, Liaoning, Shaanxi, Hubei, Tianjin, Anhui, Henan, Fujian and Hebei. The second group comprises 15 provinces, including Hunan, Heilongjiang, Jilin, Shanxi, Chongqing, Jiangxi, Guangxi, Yunnan, Gansu, Inner Mongolia, Xinjiang, Guizhou, Qinghai, Ningxia and Hainan. The annual average of R&D funds of the society at large of the former group is higher than that of the latter.

3.3.5. *Infrastructure construction*

We use the annual average of the local public expenditure on infrastructure construction of the provinces from 2003 to 2008 (unit: ¥10,000) to measure their infrastructure construction. We divide the 30 provinces into two groups based on the annual average of local public expenditure on infrastructure construction. The first group consists of 15 provinces, including, Shanghai, Guangdong, Jiangsu, Liaoning, Sichuan, Inner Mongolia, Beijing, Henan, Zhejiang, Yunnan, Xinjiang, Tianjin, Shandong, Chongqing and Anhui. The second group comprises 15 provinces, including Hebei, Hunan, Shaanxi, Guangxi, Heilongjiang, Shanxi, Fujian, Jilin, Gansu, Hubei, Jiangxi, Guizhou, Ningxia, Qinghai and Hainan. The annual average of local public expenditure on infrastructure construction of the former group is higher than that of the latter.

The GDP data are obtained from the *China Statistical Yearbook* from 2004 to 2009. The R&D fund of the society at large and the patents owned by every 10,000 people are derived from the *China Statistical Yearbook on Science and Technology* from 2004 to 2009. The FDI stock is calculated based on the data from the China Economic Information Network as well as the statistical yearbooks, almanacs of economy and statistical communiqués of each province and municipality, with reference to the calculation method of Zhang *et al.* (2004).

F-test and Hausman test were performed on each group to obtain the estimated results. Each group should be analyzed with individual fixed effects model. The regression results are shown in Table 5.

The significance of each explanatory variable in the regression results indicates that there are differences in the grouping of the five influencing factors. On this basis, we find out whether the five factors have significant influences. We discover that the significance of the outcome variables of the grouping of the former four factors has changed. Hence, that these four factors affect the influence of efficiency of R&D internationalization on its level.

However, it is impossible to determine whether there is a significant difference between the two groups divided based on local expenditure on infrastructure. Therefore, we conduct a Chow test on this factor and analyze whether there is structural stability between each two groups of samples, i.e., whether there is significant difference in the regression coefficients between each two groups of samples. Upon calculation, we have $F = 1.04 < F_{0.05}(2,112) = 3.08$. It can be seen that the structure is stable and there is no significant difference. Therefore, infrastructure construction has no significant promotional effect on the level of R&D internationalization.

According to the regression results of the 10 groups, in areas with a large market size, more foreign investment, higher degree of protection for intellectual property rights, and abundant R&D resources, technological efficiency has a significant role in promoting R&D internationalization. The impact of technological progress is not obvious. In contrast, in areas with a smaller market size, less foreign investment, lower degree of protection for intellectual property rights, and scarce R&D resources,

Table 5.　Grouped Regression Results of Individual Fixed Effects Models (2003–2008)

Indicator	Market size		Scale of foreign investment		Degree of protection for intellectual property rights		R&D resources		Level of infrastructure construction	
	High	Low	High	Low	High	Low	High	Low	High	Low
ln Tech	0.002	0.353	0.122	0.251	0.114	0.241	−0.001	0.315	0.099	0.289
	(0.075)	(3.866*)	(1.529)	(3.389*)	(1.268)	(3.276*)	(−0.037)	(3.604*)	(2.229)	(3.705*)
ln Pech	0.165	0.036	0.221	0.038	0.290	0.003	0.147	0.066	0.209	0.051
	(5.054*)	(0.333)	(2.433*)	(0.555)	(3.091*)	(0.034)	(4.096*)	(0.769)	(3.974*)	(0.584)
ln Sech	0.173	0.218	0.221	0.058	0.230	0.167	0.121	0.164	0.136	0.265
	(3.909*)	(1.313)	(2.319*)	(0.439)	(2.216*)	(1.191)	(2.736*)	(1.021)	(2.142*)	(2.128)
C	3.165	1.793	2.837	2.145	3.141	1.835	3.228	1.739	3.072	1.895
	(208.9)	(36.71*)	(108.4*)	(47.16*)	(124.4*)	(36.47*)	(193.7*)	(32.63*)	(120.6*)	(43.69*)
Adjusted R^2	0.094	0.967	0.991	0.946	0.992	0.926	0.994	0.934	0.988	0.967

technological progress plays a greater role in promoting R&D internationalization. The contribution of technical efficiency is not significant. This shows that a larger market size, more foreign investment, better protection for intellectual property rights and more abundant R&D resources make it easier for pure technological efficiency and scale efficiency to promote the level of R&D internationalization. Otherwise, the contribution of technological progress is greater. These are consistent with the measurement results for the level and efficiency of R&D internationalization in the previous sections.

Based on this, we believe that regions with a large market scale, more foreign investment, better protection for intellectual property rights and abundant R&D resources are mainly located in the eastern coastal areas that feature better economic development in China. These areas enjoy a higher level of R&D internationalization, a more mature market and a higher technological level. Their higher level of R&D internationalization is mainly affected by technological efficiency rather than technological progress for two major reasons.

First, the second-mover advantage that China's developed regions gain by imitating and using foreign technology is no longer obvious. After China started to attract foreign investment, in the first dozen years, its more open eastern regions were the first to absorb this part of investment. They imitated and utilized others' technologies and, by virtue of the resulting technological progress, enjoyed second-mover advantage that quickly brought them a higher level of R&D internationalization. In the recent years, with the improvement in technology, developed regions in China are keeping up with the developed countries. Besides, equipped with better laws for protection of intellectual property rights, the technological second-mover advantage of those regions is increasingly less obvious. The technology spillover effect has gradually spread to other less mature regions with a lower level of technology, which have to some extent benefited from the advanced technologies of the developed regions in China and the foreign MNCs.

Second, the technological efficiency of less developed regions is undermined by low efficiency of state-owned enterprises, low work efficiency of governments, and problems of the financial system (Zheng and Hu, 2005). On the contrary, in more developed regions in China, the state-owned enterprises have higher efficiency, the governments have higher work efficiency and the financial systems are better, which helps give full play to the role of technological efficiency.

4. Summary

Based on the panel data of about 30 provinces in China for the period 2003–2008, we analyze and determine the level and efficiency of R&D internationalization of each region, as well as test the factors that affect how the efficiency of R&D internationalization influences its level. The conclusions are as follows.

First, the regional level of R&D internationalization is consistent with that of economic development and the regional efficiency of R&D internationalization is positively correlated with its level. The level of R&D internationalization is higher in the eastern region and generally lower in the central/western regions. By measuring the efficiency of R&D internationalization using the DEA method, we find that regions with a higher level of R&D internationalization have a higher index for regional technological efficiency and lower index for technological progress. Regions with a lower level of R&D internationalization have a lower index for regional technological efficiency and a higher index for technological progress. This is also evidenced by the test on how efficiency of R&D internationalization influences its level.

Second, market size, scale of foreign investment, degree of protection for intellectual property rights, and R&D resources are factors that have an important impact on how efficiency affects the level of R&D internationalization. In regions with a larger market size, more foreign investment, higher degree of protection for intellectual property rights and abundant R&D resources, technological efficiency plays a significant role in promoting R&D internationalization. The impact of technological progress is not obvious. In contrast, in areas with a smaller market size, less foreign investment, lower degree of protection for intellectual property rights, and scarce R&D resources, technological progress plays a greater role in promoting R&D internationalization. The contribution of technical efficiency is not significant.

Based on the above findings, we make the following suggestions. First, the protection system for intellectual property rights should be improved. Since the direct output of R&D activities is the intangible knowledge product, the process of such activities has strong spillover effects. Therefore, compared with production activities, R&D activities have more demanding requirements on the protection for intellectual property rights in regions and countries where they are carried out. A sound protection system for intellectual property rights is an important indicator of a mature market economy and a major locational factor affecting R&D investment of MNCs. Lack of laws for protecting intellectual property rights or lax enforcement of such laws affect the investment decisions and research types of the MNCs' R&D activities in China. With the advancement of the market economic system, China is attaching more importance to the protection of intellectual property rights and has formulated some laws and regulations for such protection. However, such laws and regulations are far from perfect compared with those of the developed market economy countries. As a result, the investment of enterprises in R&D activities is affected. Therefore, China must improve the protection system for intellectual property rights and lawfully protect the rights and interests of R&D achievements of the foreign-funded enterprises, including technical secrecy, transfer of property rights, as well as prevention of counterfeiting and infringement. Meanwhile, knowledge of protection for intellectual property rights must be vigorously promoted, with focus on improving people's awareness of such protection. Second, the development of domestic R&D resources should be highlighted. As the analysis results reveal, domestic R&D resources have a significant impact on both efficiency and level. An important motivation for MNCs to engage in overseas R&D is to obtain the benefits from local technology spillovers. It is easier for regions with better innovation systems and a higher level of R&D marketization to attract overseas R&D. Therefore, local governments should strengthen the building of regional innovation systems to create a sound soft environment that attracts R&D institutions from MNCs. In the modern economy, a certain amount of technology-based infrastructure is critical in attracting R&D investment from MNCs. Investment in the government-led, market-oriented, technology-based investment in research infrastructure will not only make domestic private investment more profitable but also increase the efficiency of domestic absorption of foreign technology. The availability of skilled labor, especially an ample number of excellent scientists and engineers, is a major determinant of the R&D location of the MNCs. Governments should speed up the reform of the education system and promote the internationalization of colleges and universities. At the same time, it should strengthen the building of research institutes and highlight the diversification of their functions. Besides, they should adopt special policies and measures to support young researchers, strengthen science education and training, attract international professionals, and strengthen the training of knowledge workers through a variety of channels.

Chapter 3

Measurement of Overall Indigenous Innovation Capacity of China

Since reform and opening-up, China has accomplished remarkable economic achievements and the standard of living of the people has been greatly improved. Behind the economic development, indigenous innovation has become an important source of economic growth. A country may rely on its own capabilities to promote technological progress through technological development and innovation. It may also improve its technological level by purchasing advanced technologies and introducing foreign capital to obtain technology spillovers. American economist Soro believes that, in addition to capital and labor, which are the two basic factors of economic growth, technological progress is the other factor that boosts the economy. Soro assumes that technological progress and capital are two independent factors. The increase in capital and labor does not necessarily promote technological progress. However, the endogenous growth theory negates the premise in Solo's growth theory that technology has nothing to do with capital and labor. It points out that technological progress is determined by the internal factors of the economy. The accumulation of capital can produce a "learning effect", which leads to technological progress in the broad sense and results in spillover effects that boost economic growth.

Therefore, indigenous innovation is playing an increasingly important role in boosting economic growth, safeguarding national security, and promoting social progress. It is emerging as a decisive factor for a country or region to gain international competitive advantage for its economy. As a developing country with a relatively sound economic structure, China must grasp the core technologies of strategic industries that are related to national security and economic lifeline for economic, political, military and defense reasons. China should duly upgrade the industries, which is the key to improving its innovation capacity. The *Proposal from the Central Committee of the Communist Party of China on Formulating the 11th Five-Year Plan for National Economy and Social Development* accentuates that the indigenous innovation capacity should be strengthened and that "we must implement the strategy of rejuvenating the country through science and education and the strategy of strengthening the country through talent, take the enhancement of indigenous innovation capacity as strategic basis of a science and technology as well as the central part of the industrial restructuring and transformation of the growth model, as well as vigorously improve the original innovation capability, and integrate innovation capacity and the capabilities of introduction, digestion, absorption and re-innovation". The Party Central Committee and the State Council have placed the country's indigenous innovation capacity in a prominent position in economic and social development. Therefore, accelerating the improvement in China's indigenous innovation capacity is of great significance to addressing the challenges brought about by the new global technological and industrial revolution, building a well-off society in an all-round way, implementing the "Scientific Outlook on Development", and promoting China's socialist modernization, so that China can follow a path of indigenous innovation that features high technological content, good economic returns, low resource consumption, low environmental pollution and full play of human resources.

Since the Fifth Plenary Session of the 16th CPC Central Committee deliberated and adopted the *Proposal from the Central Committee of the Communist Party of China on Formulating the 11th Five-Year Plan for National Economy and Social Development*, many perspectives have emerged for measuring indigenous innovation capacity. However, it has been difficult to truly measure the indigenous innovation capacity of a country or a region.

1. Clarification of the Connotation of Indigenous Innovation

To address this problem, it is necessary to clarify the connotation of "indigenous innovation". According to Schumpeter (1934), innovation means establishing a new production function or supply function and introducing a new combination of production factors and conditions into the production system. Some foreign scholars have studied and explored innovation capacity from the technical and institutional perspectives. For instance, Enos (1962), Lynn (1984), Freeman (1973), and the National Science Foundation (NSF, 1976) have all defined "technological innovation". Musser sorts out and analyzes more than 300 papers on "technological innovation" and defines it as a meaningful non-continuous event characterized by its novelty and successful realization. New institutional economists such as Coase and North have conducted in-depth research on institutional innovation. American scholars Nielsen and Johnson combine the two and believe that "institutional innovation theory and technological innovation theory are mutually promoting; now technological innovation theory takes institutional innovation more seriously and is using institutional concepts with a broader and more complex approach".

In order to explain how indigenous innovation becomes the opposite of learning and imitative innovation, we follow Schumpeter's definition of innovation and redefine indigenous innovation as the process where an economy inputs its own human capital, R&D and other innovative factors and, in a certain context of innovation, independently creates new knowledge, technologies, management concepts and other innovation achievements with independent intellectual property rights, and transforms these innovations into reproduction factors.

For the above definition, we have the following interpretation. First, the actor of indigenous innovation is "an economy". Hence, there are macroscopic, mesoscopic and microscopic actors, reflecting the innovation behaviors at three levels, namely, the country, the regions (industries) and the enterprises. Second, a series of conditions such as state fiscal support is imperative for the input of innovation factors like human capital in order to yield output. Third, the output of indigenous innovation must be based on independent R&D and production without relying on external forces. Fourth, enterprises must have independent intellectual property rights for the fruits of innovation, that is, independent property rights. Fifth, the process of indigenous innovation also includes the transformation of innovation achievements into reproduction factors, that is, investing the innovation achievements from R&D into production for marketization so that they become new production factors for reproduction. Last, indigenous innovation is an output process where innovation factors are invested for innovation achievements.

Therefore, we can assume that indigenous innovation is the opposite of imitative innovation. On this basis, we believe that, throughout the technology spillover of FDI, the effect on indigenous innovation is the opposite of that on imitative innovation.

2. Measures of Indigenous Innovation

With regard to the extensive connotation of indigenous innovation, we adopt the four first-level indicators[1] and 30 second-level indicators in the *Report on the Indigenous Innovation Capacity of Chinese Enterprises* released by the China Economic Monitoring & Analysis Center of the National Bureau of Statistics in 2006 to measure indigenous innovation capacity (see Table 1).

Using the four first-level indicators and 30 second-level indicators, we use SPSS and factor analysis to determine the indigenous innovation capacity of each region.

Factor analysis is a multivariate statistical method that examines the relationships between multiple variables. It is a tool to find out how to interpret the internal structure of multiple variables with a few principal factors. In other words, a few principal factors are derived from the original variables; they retain as much information of the original variables as possible and are not correlated to each other.

Therefore, we first analyze the variance of indigenous innovation indicators of the provinces and cities in 2008[2] and obtain the total variance values, as shown in Table 2.

As Table 2 indicates, only the first three eigenvalues are greater than 1. So, they can be used as the principal factors of the above indigenous innovation indicators. Their accumulative variance contribution rate is 93.2%, which satisfies the standard for determining principal components in principal component analysis (accumulative variance contribution rate >85%). This is also evidenced in Figure 1.

---

[1] The four first-level indicators are indicators of evaluation of activities of technological innovation, indicators of environment for technological innovation, indicators of potential resources for technological innovation and indicators of output capacity of technological innovation.

[2] The FDI data for Tibet are not complete. In this book, we use 30 provinces, autonomous regions and municipalities except the Tibet Autonomous Region.

Table 1.   Indigenous Innovation Capacity Evaluation Indicator System

No.	Indicator	Unit	No.	Indicator	Unit
	Indicators of evaluation of activities of technological innovation	—	15	Percentage of national total	%
1	Number of staff for technological activities	10,000	16	Value-added rate of high-tech enterprises above designated size	%
2	Scientists and engineers	10,000	17	Ratio of high-tech enterprises to industrial enterprises in value added	%
3	Number of staff for technological activities per ten thousand people	—	18	Import and export value of high-tech products	$100 million
4	Number of R&D professionals	10,000 per year	19	Percentage of national total	%
5	R&D scientists and engineers	10,000 per year	20	Import value of high-tech products	$100 million
6	S&T expenditure	¥100 million	21	Percentage of national total	%
7	Ratio of S&T expenditure to GDP	%	22	Export value of high-tech products	$100 million
8	R&D fund	¥100 million	23	Percentage of national total	%
9	Ratio of R&D fund to GDP	%		Indicators of output capacity of technological innovation	—
	Indicators of environment for technological innovation	—	24	Number of patent applications accepted	—
10	Local government S&T appropriation	¥100 million	25	Number of invention patent applications accepted	—
11	Percentage of local public finance	%	26	Number of patent granted	—
	Indicators of potential resources for technological innovation	—	27	Number of invention patents granted	—
12	Output value of high-tech enterprises above designated size	¥100 million	28	Number of S&T papers published on the domestic Chinese periodicals	—
13	Percentage of national total	%	29	Number of contracts reached in the technology market	—
14	Value added of high-tech enterprises above designated size	¥100 million	30	Value of contracts signed in the technology market	¥100 million

Table 2.　Total Variance Explained

Component	Initial eigenvalues			Rotation sums of squared loadings		
	Total	% of variance	Cumulative (%)	Total	% of variance	Cumulative (%)
1	22.44	74.81	74.81	13.39	44.63	44.63
2	4.90	16.33	91.14	9.62	32.07	76.70
3	1.18	3.93	95.07	5.51	18.37	95.07
4	0.56	1.87	96.94	—	—	—
5	0.35	1.16	98.10	—	—	—
6	0.19	0.65	98.74	—	—	—
7	0.12	0.41	99.15	—	—	—
8	0.08	0.27	99.42	—	—	—
9	0.07	0.22	99.64	—	—	—
10	0.03	0.10	99.74	—	—	—
11	0.02	0.08	99.82	—	—	—
12	0.02	0.06	99.88	—	—	—
13	0.02	0.05	99.93	—	—	—
14	0.01	0.03	99.96	—	—	—
15	0.01	0.02	99.98	—	—	—
16	0.00	0.01	99.99	—	—	—
17	0.00	0.01	99.99	—	—	—
18	0.00	0.00	100.00	—	—	—
19	0.00	0.00	100.00	—	—	—
20	0.00	0.00	100.00	—	—	—
21	0.00	0.00	100.00	—	—	—
22	0.00	0.00	100.00	—	—	—
23	0.00	0.00	100.00	—	—	—
24	0.00	0.00	100.00	—	—	—
25	0.00	0.00	100.00	—	—	—
26	0.00	0.00	100.00	—	—	—
27	0.00	0.00	100.00	—	—	—
28	0.00	0.00	100.00	—	—	—
29	0.00	0.00	100.00	—	—	—
30	0.00	0.00	100.00	—	—	—

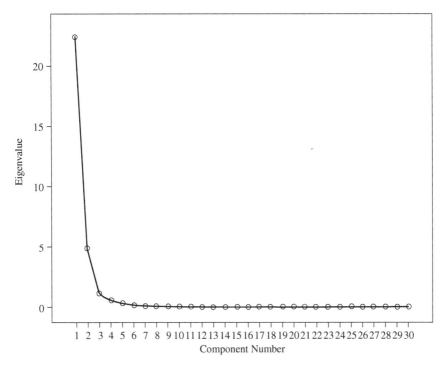

Figure 1.   Common Factor Scree Plot.

As Figure 1 reveals, the eigenvalues of the first three principal components change significantly while those of the fourth onwards tend to be balanced. This shows that extracting three common factors can have a significant effect on the information description of the original 30 indicators. Therefore, these three principal factors are sufficient to explain the indigenous innovation capability of a region. In this case, the three factors are denoted as F_1, F_2 and F_3. The loaded components of these three principal factors have been detailed in Table 3.

As Table 3 indicates, F_1 is loaded with more information about high technology.[3] Therefore, it can be considered that F_1 represents the high-tech capability. Similarly, F_2 can be regarded as the input of indigenous innovation capability and F_3 the macro-environment.

---

[3] The 12th to 23rd indicators include over 90% of information.

Table 3.   Principal Component Load Matrix

Indicator	Principal component		
	1	2	3
Staff in technological activities	0.50	0.76	0.42
Scientists and engineers	0.49	0.50	0.71
Staff in technological activities per ten thousand people	0.09	0.98	0.14
R&D professionals	0.51	0.69	0.50
R&D scientists and engineers	0.53	0.64	0.53
S&T expenditure	0.50	0.65	0.54
Ratio of S&T expenditure to GDP	0.03	0.93	0.19
R&D Funds	0.49	0.61	0.61
Ratio of R&D Funds to GDP	0.04	0.95	0.22
Local government S&T appropriation	0.67	0.85	0.42
Percentage of local public financial expenditures	0.43	0.80	0.24
Percentage of output value of high-tech enterprises above designated size to national total	0.92	0.20	0.34
Percentage of value-add of high-tech enterprises above designated size to national total	0.89	0.15	0.41
Value-added rate of high-tech enterprises above designated size	−0.050	−0.57	−0.32
Ratio of high-tech enterprises to industrial enterprises in value-add	0.52	0.74	0.09
Percentage of import and export value of high-tech products to national total	0.96	0.0	0.19
Percentage of import value of high-tech products to national total	0.95	0.24	0.17
Percentage of export value of high-tech products to national total	0.96	0.16	0.21
Number of patent applications accepted	0.61	0.20	0.75
Number of invention patent applications accepted	0.43	0.52	0.73
Number of patent applications granted	0.59	0.18	0.73
Number of invention patent applications granted	0.50	0.40	0.75
Number of S&T papers published on the domestic Chinese periodicals	0.26	0.70	0.60
Number of contracts signed in the technology market	0.24	0.30	0.90
Value of contracts signed in the technology market	0.06	0.12	0.95

Then, based on the variance contribution of the three principal factors in Table 2, the comprehensive statistic of indigenous innovation F is constructed:

$$F = \frac{\lambda_1}{\lambda_1 + \lambda_2 + \lambda_3} F_1 + \frac{\lambda_2}{\lambda_1 + \lambda_2 + \lambda_3} F_2 + \frac{\lambda_3}{\lambda_1 + \lambda_2 + \lambda_3} F_3. \qquad (1)$$

From Eq. (1), the F of each province and municipality in 2008 can be derived. Standardize F and the comprehensive indicator of each province and municipality I can be obtained.

3. Measurement of Regional Indigenous Innovation Capability

Based on the above method, the comprehensive indicators of indigenous innovation capability of the provinces and municipalities from 2000 to 2008 are obtained, as shown in Table 4.[4]

As Table 4 reveals, Guangdong, Beijing, Jiangsu and Shanghai, all of which have attracted more foreign investment, show stronger indigenous innovation capacity than other regions.

4. Evolutionary Analysis of Regional Indigenous Innovation Capacity

Based on the calculation of regional comprehensive capacity of indigenous innovation in the previous section, we figure out the nine regions with the highest arithmetic mean of regional indigenous innovation capacity from 2000 to 2008. They are Guangdong, Beijing, Jiangsu, Shanghai, Tianjin, Zhejiang, Shandong, Liaoning and Sichuan. All their average values of comprehensive indigenous innovation capacity exceed 100, as shown in Figure 2. It is evident that, except Shaanxi, the rest are all eastern coastal regions in China. The comprehensive level of Guangdong, Beijing, Jiangsu and Shanghai even surpasses 150. Similarly, Guangdong,

---

[4]Statistics on indicators of Inner Mongolia, Guangxi and Qinghai before 2000 are not complete, which prevents factor analysis. To guarantee the continuity and authenticity of data, this book focuses on the period, 2000–2008.

Table 4.   Indigenous Innovation Capability of Provinces and Municipalities (2000–2008)

	2000	2001	2002	2003	2004	2005	2006	2007	2008
Beijing	226.38	194.29	177.73	187.18	185.52	197.20	192.66	194.83	194.17
Shaanxi	104.66	102.95	105.36	99.96	96.98	94.51	103.30	97.71	91.95
Hebei	79.81	86.08	86.44	82.65	79.66	76.66	77.00	86.35	87.51
Sichuan	104.10	102.76	100.63	98.98	94.08	92.10	99.86	111.95	116.19
Liaoning	108.65	111.04	113.17	108.13	106.87	101.15	109.72	106.22	101.96
Jiangxi	73.68	77.77	78.91	77.33	75.43	75.47	77.95	79.31	88.34
Hainan	65.66	70.27	70.27	69.20	68.94	68.88	73.25	67.93	67.39
Heilongjiang	77.71	87.00	88.10	84.28	80.90	80.06	78.05	85.92	86.65
Shanghai	167.85	157.06	151.35	168.12	179.42	190.99	173.85	168.17	166.58
Jilin	81.20	83.17	83.97	81.29	79.58	79.62	85.43	81.65	86.47
Yunnan	70.38	76.70	77.01	73.60	71.92	69.95	69.05	72.60	74.11
Anhui	71.90	80.26	81.09	79.08	77.22	76.40	75.54	81.22	80.99
Chongqing	71.62	79.86	80.73	80.83	79.64	80.04	81.45	85.57	86.97
Jiangsu	167.71	154.38	155.29	171.78	185.46	192.28	174.70	178.56	177.65
Qinghai	59.19	68.78	69.92	68.69	66.24	65.62	65.37	67.62	68.93
Hubei	90.09	97.69	97.32	95.05	89.00	90.02	95.96	87.12	92.56
Henan	83.69	86.70	86.54	82.66	81.42	78.31	82.24	81.44	82.58
Hunan	80.94	88.21	89.07	83.71	82.17	81.33	88.48	86.62	86.05
Guangdong	306.01	245.03	240.88	258.11	265.93	260.08	260.17	265.13	267.46
Gansu	70.04	76.01	76.42	74.61	71.74	71.35	73.47	81.58	83.91
Guizhou	70.64	75.03	76.03	75.20	73.69	72.84	77.34	80.74	79.19
Zhejiang	104.99	111.26	116.10	113.92	120.06	120.07	112.84	126.56	120.29
Shanxi	70.58	76.70	77.97	75.22	74.51	72.52	75.97	79.96	78.39
Guangxi	70.82	74.79	75.71	75.02	72.82	71.94	76.20	71.65	72.55
Xinjiang	61.76	70.52	71.21	68.75	67.72	66.63	71.59	70.82	71.55
Shandong	109.04	113.56	115.21	111.80	112.36	109.97	115.69	115.16	112.99
Fujian	103.18	96.98	98.46	99.48	100.00	99.07	97.03	98.27	95
Ningxia	64.04	69.68	71.27	69.56	68.43	68.74	69.32	67.10	66.22
Inner Mongolia	64.10	71.70	73.14	71.20	69.71	69.28	74.26	65.48	71.12
Tianjin	119.59	113.76	114.25	114.61	122.61	126.96	118.28	119.91	117.22

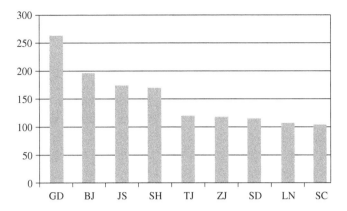

Figure 2. Nine Regions with Comprehensive Indigenous Innovation Capacity Higher than 100 (2000–2008).

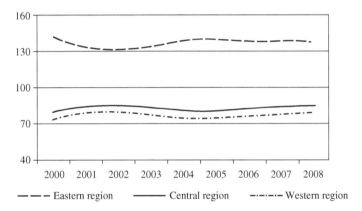

Figure 3. Evolution of Indigenous Innovation of Eastern, Central and Western Regions (2000–2008).

Jiangsu and Shanghai also top the rankings in terms of comprehensive indigenous innovation capacity.

However, apart from these nine regions whose comprehensive level is higher than 100, only eight regions have a comprehensive level between 80 and 100 and the others around 70. It can be seen that there is a huge difference in the indigenous innovation capacity across regions.

Viewed in terms of economic regions, namely, the eastern, central and western regions, the regional comprehensive capacity of indigenous innovation shows distinctive regional characteristics. As Figure 3 shows, from

2000 to 2008, the level of indigenous innovation in the eastern region has remained above 130, while that in the central region is close to that in the western regions, at around 80.

The eastern region is distinctively different from the central and western regions with respect to indigenous innovation capacity. In terms of patterns of evolution, all the three regions show changes in an S-shaped curve. However, the changes are small. For the eastern region, the biggest gap in the period is found between 2005, at 140, and 2002, at 130. Yet, the gap is actually quite small. In addition, the evolution trend of the eastern region is basically the opposite of those of the central and western regions. In the interval, while the level of indigenous innovation of the eastern region increases, that of the central and western regions steadily declines. The regional comprehensive capacity of indigenous innovation from 2000 to 2008 approximately exhibits a vase shape. This is because, on the one hand, China's entire level of R&D investment has remained stable for a long time. There have not been major fluctuations as GDP grows. Therefore, when R&D investment is given, R&D expenses of different regions are significantly exclusive. On the other hand, the subjects of R&D in China are not enterprises at the micro level, but research institutes as well as colleges and universities. Hence, for R&D expenditure, more consideration is given to fairness than efficiency. As a result, there is exclusivity in the regional R&D expenses.

Chapter 4

The Facilitating Effect of R&D Internationalization on Indigenous Innovation in China

1. Technology Spillover Effects of R&D Internationalization

1.1. *Definition of technology spillover*

The World Intellectual Property Organization defines technology as part of knowledge and believes that it is the systematic knowledge for producing a product or a service. This type of knowledge can include an invention of a product or process, a form design, a utility model, a new variety of plant or animal, or a professional skill in design, layout, maintenance and management.

Helleiner (1975) interprets technology in the sense of patent and believes that technology includes legally recognized patents and trademarks, technology that is not legally patented, proprietary knowledge that is not patented as well as experience contained in skilled work and technology embedded in tangible goods.

Hayden (1976) argues that technology is knowledge in terms of management, organization, and skills necessary for the production of a particular industrial product. This definition means that technology is a production input that includes both processes and skills. Processes include knowledge about how to produce a product, including data on plant construction and design, equipment requirements, etc. and information such as raw material standards. Skills involve the ability to transform professional technology into market products, including general industrial skills required to produce a product, as well as the ability to solve the problems of production and maintenance, plan production process and ensure product quality.

Porter (1985) asserts that technology is an umbrella term that covers every part of the corporate value chain. If any single standard is used to measure technology, bias is likely to emerge because technology includes not only objective hardware evaluation but also software technology like business management.

Brooke (1988) divides technology into three levels: The first level is a relatively broad range of functional technologies, the second level involves technologies in terms of product processes and the third level includes proprietary knowledge in the true sense.

Robinson (1990) analyzes the horizontal composition of technologies, including indivisible core technologies or peripheral technologies, user rights or licensing of knowledge (especially technology licensing), hardware, typical equipment containing technology or other intermediate products and final products and software. Documents and verbal knowledge discussions, for example, belong to non-materialized technologies. He also makes a vertical classification of technologies, including user technologies (techniques related to the use of technology), product absorption technologies (the techniques of using a technology under specific conditions), production technologies (techniques required for copying technology), and design technologies (such as techniques required to design a product). In short, technology generally refers to people's effective means of using various rare resources to engage in various economic activities. Its extension covers various aspects, such as product, manpower and organization (Li, 1999).

The definition adopted in this book is more in line with Robinson (1990), i.e., the technology spillovers caused by the investment of the MNCs in China should include the relevant software and hardware technologies of the products as well as the professionals stationed in China by MNCs and the improvement in the managerial skills brought about by the training of local professionals. According to Hughs (1994), technology cannot be understood as a tool or a path of learning because it has a great impact on the changes of human society. It should also include many social factors, such as value, system, interest groups, social classes and political and economic forces, which together constitute a technological system, i.e., a technological system is used in the place of the traditional definition of technology. Outside the technological system, there is the environment. Technological system and the environment shape and influence each other. In this book, we adopt Hughs's point of view to study the technology spillovers caused by the direct investment of MNCs in China because the MNCs cannot avoid interacting with the internal environment of China.

Technology spillovers are a part of the globalization of technological innovation. Therefore, before defining technology spillovers, we first classify the manifestations of globalization of technological innovation. According to Archibugi and Lammarino (1999), the connotation of globalization of technological innovation has three aspects: the use of technology produced by a single country around the world, the globalization of technological innovation, and the global technological cooperation. The three aspects support rather than repel each other. They can be shown at the enterprise level as well as at the national level, as indicated in Table 1. Technology spillovers emerge in the context of globalization of technological innovation.

There are many definitions of technology spillover. In this book, technology spillover is mainly an involuntary form of technology transfer, in the sense that MNCs setting up subsidiaries in host countries will lead to

Table 1.   Forms and Subjects of Globalization of Technological Innovation

Type	Subject	Form
Global use of domestic technological innovation	Enterprises and individuals in pursuit of profit	Export of innovative technological products — License and transfer of technologies and patents — Overseas production of innovative technological products
Globalization of technological innovation	MNCs	R&D and technological innovation activities (R&D internationalization) at home and around the world
Global S&T cooperation	Universities, national research institutes, companies and MNCs of different countries	Cooperation on S&T projects, personnel exchanges, joint ventures, etc.

local progresses in technology or productivity, but their subsidiaries cannot obtain all the income (Chen, 2001). It also refers to the case when developed countries make direct investment in developing countries, their advanced technology, management philosophy, managerial experience, etc. can be transferred to other enterprises in host countries through certain channels, thus influencing the economic development and growth of host countries (Li and Su, 2001). Technology spillover is also a comprehensive dynamic process. There is transfer through hardware technologies like imported machinery and equipment as well as intermediate products. There is also transfer and diffusion of software technologies, such as technical service consulting, technical personnel training, organizational and managerial skills, and development of entrepreneurship. More importantly, there is the dynamic education on technological development and innovation mechanism (Xu, 2002).

Technology spillover refers to the process in which production technologies imported from abroad spread from enterprises that introduce the technology to other departments of the industry, suppliers of raw materials for the said enterprises and other departments of related industries (He *et al.*, 1996). When industrial technicians leave the enterprises that introduce technologies and work in other enterprises, there will always be some kind of technology diffusion. In addition, enterprises that introduce technologies can also help other companies in the industry to obtain technologies that they have introduced, or they can provide technological services to local raw material suppliers and partner enterprises, so as to localize raw materials, parts, assets and equipment. These can all be regarded as technology spillover effects. Today, technology spillovers are no longer limited to the field of production. They include not only technical knowledge but also business management knowledge, such as financing, marketing, and management. The forms of technology are diversified. Thanks to new forms of technology, technology spillovers are no longer a replication process in different regions. Since the transfer of organizational technologies is influenced by the economic, social and cultural backgrounds of different host countries, technology transfer can happen in different forms at different levels (Xie and Zheng, 2001).

Vernon (1966) describes the occurrence of technology transfer based on product cycle. He classifies the development of products into new product period, product maturity period, and standardization period based on the technological level. In the new product period, advanced countries are engaged in R&D of new products to meet local market demand. In this case, price elasticity is small and there is less competition. When products are promoted, demand increases, production scale expands, production costs reduce, and competitors also appear. In the case of high product isomorphism, manufacturers use price as a tool for competition, which forces manufacturers to reduce production costs and look for a more favorable investment location overseas. Technology is then transferred abroad as part of overseas investment.

Komoda (1986) argues that technology spillovers should involve a shift of the ability to understand and develop the technologies introduced (Li, 1999). The standard for judging whether technology diffusion is successful is that when technology is introduced without external help, technology can be absorbed, manipulated and repaired, and the introducing enterprises are able to improve, spread and develop the introduced technology. As this definition indicates, technology diffusion is not simply the acquisition of production technology, but the development of the technological capabilities of the introducing parties.

Wang and Blomstrom (1992) divide technology spillovers into convergency spillovers and competition spillovers. Convergency spillovers refer to the tendency that the technological level of enterprises in host countries approaches or converges with that of MNCs because of learning and imitation. Competition spillovers imply that the entry of MNCs increases competition and forces the enterprises in host countries to use higher technologies to narrow their technological gap and prompt the MNCs to improve the level of transferred technology to maintain their strengths. As a result, technologies of host countries improve on the whole.

Ozawa (1992) argues that five structural features of the world economy significantly affect the cross-border flow of various factors, including technology:

(1) There are differences between the supply side and the demand side within each economic entity. The supply side refers to the level of factor endowment and technological capability while the demand side mainly involves factors, such as consumer demand and consumption preferences and habits. In the real economy, it is often difficult to match the two. It is a common phenomenon that there is a gap between the supply side and the demand side of economy within a country.
(2) Enterprises are the creators and traders of various kinds of intangible assets. Every enterprise can be seen as the main creator and promoter of technology and sales channels.

(3) The hierarchical structure of economic development level and strength of all countries is obvious. According to the growth strength of all countries, the world economy is divided into the leading economic entities at the center of growth and the followers of the economy. In this sense, the world economy also has obvious regional hierarchical structural features. In other words, countries are in different stages of structural upgrading, with different levels of per capita income. When measured by technological capability and factor proportion, countries are in different stages of dynamic comparative advantage.

(4) Each country has its own stages of natural evolution in terms of the economic structural upgrading and development. A well-developed economy is achieved through multistage industrial upgrading, each stage with corresponding factor endowment (capital–labor ratio) and technological capability. Completing one stage involves a historical process. A country advances by beginning with the labor-intensive initial stage, then running through the manufacturing stage with simple technologies, then moving on to the relatively material-intensive follow-up stage, and finally to the developed stage with human capital-intensive growth. Therefore, the industrial structure of a country progresses in an orderly and gradual manner.

(5) The foreign trade and foreign investment policies of various countries show a very strong trend of shifting from inward to outward. Meanwhile, there is a growing consensus that the government can play an active role in the increasingly expansive market system. The background is that developing countries that implement export-oriented strategies are better in terms of economic growth and structural upgrading than those adopting import substitution strategies. In the competitive market, the government offers active support and assistance to both export-oriented industrial development and structural upgrading.

Technology spillover effects emerge mainly because knowledge and technology have the characteristics of public property. This is also the reason why many countries take the initiative to attract investment from MNCs. Scholars believe that MNCs are the owners of advanced technologies and skills as well as the primary agents of technology transfer. Investment and production activities of MNCs generate pro-competitive effects, externalities and technology spillovers in host countries. Therefore, the entry and existence of MNCs have greatly facilitated the technological growth of host countries (Caves, 1974; Findlay, 1978). The technology spillover effects of MNCs, on the one hand, are related to the technological gap and reflected in factor endowment as well as labor productivity, and on the other hand, depend on the market environment, the game results of decision optimization of MNCs and the enterprises in host countries (Wu and Wu, 2002). Although technology spillovers of direct investment by MNCs are beneficial to the industries and economic development of host countries, they have received criticism from some scholars from the perspectives of economic nationalism and dependence theory.

1.2. *Technology spillover effects*

Technology spillover effects are the major motivation for governments and enterprises of host countries to attract R&D investment by MNCs. Technology spillover effects on the local enterprises of host countries are the direct causes of technological progress. Given the externalities of knowledge, after MNCs set up R&D institutions in host countries, they cannot avoid generating technology spillover effects on host countries. UNCTAD believes that the technological link and learning effect between MNCs and enterprises in host countries are mainly reflected in technology transfer, technology diffusion and technology creation. Among them, technology transfer is mainly mirrored by the technological relationship between the internal headquarters of MNCs and their subsidiaries. Technology diffusion is mainly manifested as the technological connection between MNCs and the local enterprises of host countries. The creation of technology is shown as the influence of MNCs on the innovation capacity of host countries (UNCTAD, 2000).

Firstly, the R&D institutions of MNCs generate technology spillovers by connecting with rivals in the same industry, upstream and downstream affiliates and other non-profit R&D institutions. From the perspective of industrial structure, technology spillover effects fall into two broad categories. The first category is technology spillover effects within the industry. The existence of MNCs enhances the competition in the markets of host countries and also improves their efficiency in resource allocation, which benefits local competitors. Meanwhile, local enterprises also improve their own technology and management level through learning and imitation during competition. The second category is interindustry technology spillover effects. MNCs have an impact on enterprises engaged in the vertical division of labor and improve their technological level, for example, by providing suppliers and users with technological guidance, training and management consulting.

Secondly, for competitors in the same industry, the entry of the R&D institutions of the MNCs will intensify market competition. In the competitive market structure, the technology spillover effects of MNCs tend to be greater. For one thing, as the circulation and renewal of commodities speed up, the R&D and application of new technologies accelerate, so does the speed of technology dissemination and diffusion. For another, circulation of commodities increases the opportunities for enterprises in host countries for imitation and learning, which are essential means for host countries to improve their technological level. Fu (1999) believes that any technologies, including complex ones, can always be learned. This is also the reason why imitation is possible. All products in the world can be imitated and all enterprises (including those with strong R&D) imitate others' products. Moreover, compared with innovation, imitation enjoys advantages, such as requiring less labor and investment, involving lower risk, and being more easily accessible. Of course, many companies imitate innovators' products through reverse engineering to better avoid disputes over intellectual property rights, thus entering the stage of imitative innovation.

For upstream and downstream affiliates, MNCs set up R&D institutions in host countries largely to provide technological services to local enterprises that have economic links with them. In the network with MNCs as the center, R&D Institutions are actually serving the entire network. Enterprises of the host countries participating in the network will certainly benefit from it. For instance, the production technologies of original equipment manufacturers and the quality of goods from suppliers in host countries will be improved, new products will be designed for local markets and technological services will be provided to local clients. In particular, in the context of economic globalization, MNCs have to redefine their operating institutions to adapt to the needs of technological development and market changes. Under the trend of refocusing on core business operations, MNCs and host countries maintain closer cooperation and sharing of technology. As a result, the affiliates obtain a higher level of technology spillovers.

Thirdly, R&D institutions of MNCs usually have various kinds of technological contact with non-profit R&D institutions in host countries, cooperate with them on technology or share data with them, outsource

R&D activities of non-core projects, and undertake some research work. Since R&D institutions of the MNCs have experience in R&D project management and system building, these non-profit R&D institutions can increase their internationalization and R&D level of projects through various forms of cooperation and gradually equip themselves with the ability to have dialogues with MNCs and join their technology alliances.

With the above three channels of technology spillovers considered, this book classifies four categories of technology spillover effects, as shown in Figure 1.

1.2.1. *Human capital effect*

Human capital is an umbrella term that covers technological personnel engaged in R&D and processes, as well as technicians, skilled workers

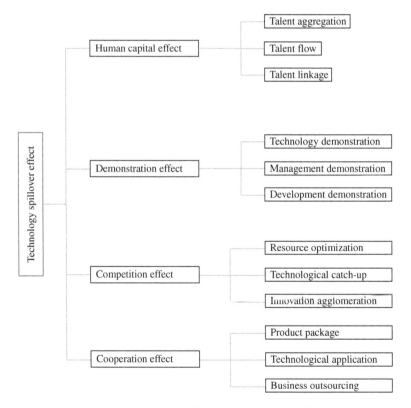

Figure 1.   Categories of Technology Spillover Effects.

and managers. As known by all, human beings are the most active and revolutionary factor. The value of competitive advantages of any company can only be fulfilled only when they are made use of by people. The same holds true for technology, whose value cannot be completely fulfilled by equipment and technologies. Therefore, the effective operation of the investment projects of the MNCs in host countries must be integrated with the development of local human resources. MNCs invest a large amount of funds on training for the localization of professionals and send technology training experts from parent companies to carry out special training sessions. The technological and managerial skills and knowledge needed by MNCs are continuously accumulated, improved and innovated via human resources. In addition, human capital flows. Labor mobility primarily means the flow of researchers, engineers, and skilled workers. When those workers flow in the market or start their own business, the knowledge and skills they have accumulated will spill over to their next jobs and they may even become competitors in the same industry. At the same time, with the increase in human capital, the capability of the workforce of host countries is improved. Skilled workforce quickly emerges, lowering the costs on technology transfer for MNCs and facilitating the transfer of higher-level technologies to host countries. In this way, a new round of technology spillovers appears and a positive feedback amplification mechanism is formed. To sum up, the human capital effect is reflected in the following three aspects:

(1) **Talent aggregation:** When MNCs are setting up R&D institutions in host countries, they firstly make full use of the local technical talent. They generally focus on promoting the localization of R&D professionals, most of whom are college graduates recruited in China or overseas returnees introduced. Meanwhile, a major feature in the selection of location for a new R&D institution is the density of development-oriented technological talent. This strengthens the aggregation of talent in the industry. According to statistics, for the branch institutions of MNCs in China, local professionals and overseas returnees take up more than 95% of their staff. Such obvious talent aggregation effect provides the possibility that host enterprises make technological progress.

(2) **Talent flow:** In the early days, the jobs provided by MNCs were highly challenging, which allowed the talent to give full play to their capabilities. Besides, the working environment and salary of these

positions were superior. As a result, the talent flow tended to be one way. However, with the economic development of host countries and the growth of local enterprises, the flow of talent between MNCs and local enterprises becomes inevitable. Through such a flow, the technology and managerial knowledge of MNCs begin to spill over. South Korea, Thailand, Malaysia, and Taiwan Province of China have successfully upgraded the technological capabilities of local companies by introducing technologies. The flow of talent between foreign-funded and domestic enterprises offers important experience.

(3) **Talent linkage:** Since the products, technologies and management of MNCs reflect the development trend of the industry, the direction and content of internal talent training sessions are often closely related to their strategic interests and development situation. The content and methods of MNC training sessions for internal employees can be directly used or referred to by the corresponding enterprises in host countries for training staff. Meanwhile, MNCs usually offer special training sessions for staff of the stakeholders and have technical communications with them, which also greatly improves the human capital structure of enterprises in host countries and facilitates the refinement, renewal and upgrading of the technological professionals' knowledge. Hence, this also generates significant technology spill-over effects.

However, to guarantee their own competitive advantages in technology, MNCs minimize the technology spillovers through technology lock-in strategies and other methods. This is a relatively negative aspect.

1.2.2. *Demonstration effect*

The major reason why MNCs can compete with local companies in host countries is that they have proprietary technologies, knowledge of business management, experience in marketing development, and a complete information system. In the era of knowledge economy, the close relationship between information flow and technology transfer has been further strengthened via the internet. This may continue to deepen. Information technology is changing the economic structure, ways of production and lifestyle with unprecedented influence and penetration. It will reduce barriers in terms of time, space and language, speed up technology transfer and increase the efficiency of such transfer. Information technology is applied in business operations. Information systems are built for various kinds of technologies of enterprises to make them more intact. In particular, information technology is used as a carrier of enterprises' proprietary technologies to sublimate them, which is conducive to the maintenance and development of proprietary technological achievements. Rebentisch *et al.* (1997) argues that information in the ordinary sense accounts for almost 75% of the transferred technologies while program and hardware each take up only 13% of the total transfer. These types of knowledge and information have brought valuable learning opportunities to local enterprises of host countries. In this sense, MNCs play a demonstrative role in host countries, which, specifically, includes the following three aspects:

(1) **Technology demonstration:** MNCs are often technology leaders in the industry, with an enduring driving force for innovation, profound technological development and valuable proprietary technologies, all of which guarantee their technological advantages. For local enterprises of host countries, due to the existence of industrial barriers and the risk of innovation, their desire for innovation is insufficient. As technologies are constantly evolving, access to the sources of technological innovation is the key to continuous technological progress. When MNCs establish R&D institutions in host countries, the technologies they use will have an impact on similar enterprises or technology-related enterprises in host countries in many tangible and intangible ways, such as equipment, products, personnel contact, and customer technical data. How strong the demonstration effect of technologies is depends not only on the existing technological gap between the two parties but also on the ability of the enterprises of host countries to understand and apply technologies. The technological basis of domestic enterprises in China is relatively strong, and the technology demonstration effect brought about by the entry of MNCs is more obvious in some industries.

(2) **Management demonstration:** In addition to advanced technologies, host countries also introduce managerial skills and experience in effective application of these advanced technologies. The advanced management techniques of MNCs not only allow efficient use of the advanced technologies introduced but also provide other enterprises of host countries with opportunities to see the process and learn from it so that they understand how advanced management technologies can help enterprises to improve efficiency. For example, from the

perspective of internal management, the MNCs demonstrate how to integrate the global production, marketing, and technological development networks into an efficient whole for the purpose of reducing costs and increasing market share. In terms of strategic development, they demonstrate how to seek new opportunities for expansion at low cost as well as new sources of knowledge and innovative ideas across the globe and how to effectively use various support systems. All these demonstrations greatly benefit the local enterprises of host countries by allowing them to broaden their horizons and perform imitation.

(3) **Development demonstration:** Successful development of products to a large extent depends on the grasp of the direction of global technological development and the understanding of market demand. This has been made more difficult by the exponential growth of knowledge and information. For enterprises in the developing host countries, it is no mean feat to capture global business opportunities and proactively adapt to market demand simply based on their own knowledge. For the orientation toward technological development, MNCs give full play to their understanding of technologies and the market so as to accurately grasp business opportunities and judge the trends of technological development. Local enterprises of host countries may understand, observe as well as imitate technologies and market orientations of new products from MNCs, thus rapidly increasing their own success rate for developing new products and reducing market risks accordingly.

1.2.3. Competition effects

The existence of MNCs breaks the original market equilibrium of host countries. They will push the local enterprises to improve production efficiency and continuously introduce new products and processes to safeguard their market share. Meanwhile, they also activate the local supply of technological factors. The competition effects mainly include the following three aspects:

(1) **Resource optimization:** Due to the existence of industrial barriers in host countries, the original market structure cannot be the optimal allocation of resources. The entry of MNCs into the market breaks the industrial barriers and intensifies the local market competition.

Those enterprises with lower resource efficiency lose their competitiveness in the new market environment and are eventually eliminated from the market. Enterprises that have obtained higher profits under market protection have to carry out technological innovation, R&D and optimization and reorganization of other enterprises in order to maintain the original profitability. The enterprises get rid of the original industrial protective barriers and gain new impetus for survival. The monopolistic distortion of resource allocation is thus corrected and the productivity of all enterprises in host countries is improved. In this way, the resource allocation of host countries is optimized and its efficiency is improved.

(2) **Technological catch-up:** The entry of MNCs imposes strong external pressure on local enterprises of host countries and speeds up technology diffusion and transfer. This stimulates and forces local enterprises to increase investment in technology and R&D as well as to seek and improve suitable new technologies to enhance their competitiveness. On the one hand, it is very likely that the local enterprises will quickly improve their skills through active imitation and learning. On the other hand, the technological behaviors and R&D efforts of host countries press the MNCs to adopt more advanced technologies to ensure higher profitability. As MNCs and local enterprises try to catch up and surpass each other, technological level of host countries is rapidly improved.

(3) **Innovation agglomeration:** The production of MNCs is characterized by R&D and production of high-tech products. These features determine that its industrial agglomeration is different from the traditional industrial agglomeration. The factors for technological innovation agglomeration have changed. Among the combinations of factors, intelligence intensiveness, developmental technologies, talent, information network, infrastructure, and appropriate production and living environments are very important. The intensifying competition leads to a very active supply of the above-mentioned innovation factors of host countries and plays a decisive role in the R&D investment of other MNCs in the industry.

1.2.4. *Cooperation effect*

The entry of MNCs into the market of host countries has apparent cooperation effect while bringing pressure of competition to the industry.

They work horizontally with local enterprises of host countries as well as vertically with enterprises of other industries. The cooperation effect is mainly found in the following three aspects:

(1) **Product package:** Product package consists of two levels. One is an internal–external relationship, that is, the combination of key and peripheral products of the same product. The other is a primary–secondary relationship, that is, the use of raw materials and parts of the products in the finished products. The first level usually refers to voluntary cooperation between MNCs and the local enterprises of host countries in the same industry, driven by the pressure from production costs or the industrial policies of host countries and for the purpose of expanding market share in host countries. This kind of cooperation allows both parties to obtain the given market benefits and brings products, technologies and management knowledge to local enterprises. An example is the cooperation between MAN and Shaangu on turbine compressors. The primary–secondary relationship means MNCs tend to strengthen their vertical links with local enterprises and establish a more stable relationship with the local environment. They have higher requirements on the quality, technology and performance of the raw materials and parts they purchase from host countries. They may even provide corresponding technical standards and assistance to ensure that the suppliers and the products meet the requirements of the package.

(2) **Application of technology:** This also includes two aspects. The first aspect focuses on market development. From successful R&D to market promotion, a new product takes a large amount of energy and money to guide consumers. For local enterprises, even if their products have a higher technical level and quality, it is difficult for them to become market forerunners due to the lack of brand effect and market experience. They may even thus suffer financial crisis. MNCs, however, have robust strength; they are also the leaders of consumer trends. When the market demand is activated, the entry cost of local enterprises of host countries is significantly reduced, which makes it easier for application of technology. In the other aspect of the effect of technological application, the suppliers of parts and raw materials are developed because of the entry of MNCs so that the latter have easy access to high-quality raw materials and components at a lower cost. In this way, when the design, integration and assembly of the

local enterprises of host countries reach a certain level, the final products will take the lead in terms of technology and the overall technical level of host countries industry will be improved.

(3) **Business outsourcing:** MNCs do not enter host countries to fight alone. They will make the best use of the local resources, establish various kinds of contact with local enterprises or outsource some R&D tasks to them. Through cooperation with the R&D institutions of MNCs, enterprises of host countries often gain rich and systematic technical information and experience in technical management to enhance their research capabilities. Sometimes, MNCs will also help their partners improve existing technologies for the needs of overall development of technological innovation. Although MNCs choose partners with their global business objectives considered, in any case, technological cooperation and outsourcing provide an improved technical approach for local enterprises of host countries, which has to some extent narrowed the technological gap between developing host countries and MNCs.

2. Technological Transmission Mechanism of R&D Internationalization

In the innovation process, the life cycle of technologies is constantly alternating as they evolve from old to new. For such a transformation to be completed, R&D investment plays a central role. The transmission mechanism behind how R&D investment promotes technological innovation can be explained with the four kinds of effects detailed in the following sections.

2.1. *Stimulating effect*

The stimulating effect of R&D on technological innovation can be delineated by the view of traditional theory that innovation is a linear process or by that of systematics that innovation is a nonlinear process. Some of the early products of R&D activities are new knowledge products and new scientific theories. For the construction of a model for the relationship between knowledge innovation and technological innovation, the traditional theory, i.e., the innovation chain theory, considers innovation as a linear model and a chain process that consists of basic research, applied research, development of new technologies and products, production and sales (marketization), from knowledge innovation to technological innovation. Innovation starts with basic research (knowledge innovation). The increase in scientific investment at the upstream will directly boost the output of new technologies and products at the downstream. In other words, it is the process from scientific research through knowledge innovation to technological products (technological innovation). From the viewpoint of systematics, the innovation system is a nonlinear self-organized open system far from the state of equilibrium. It has the characteristics of a system, such as openness, imbalance, irreversibility and instability. Technological innovation is the transition of innovation system from one state to another state in internal fluctuations and external disturbances. It is a complex dynamic systemic process as well as the result of the intricate and comprehensive interconnection and interaction between many participants. In this way, technological innovation is more than just a chain process. The direct correlation between technological innovation and knowledge innovation (basic research) varies greatly across industries (see Table 2). Some technological innovation processes are not necessarily the direct result of knowledge innovation. However, since knowledge innovation is fundamental, technological innovation at the beginning is

Table 2.   Direct Correlation Index Between Basic Research and Technological Progress in Related Industries Across Disciplines

Disciplines where technological progress is made		Correlation index to basic research
Basic disciplines	Biology	14
	Chemistry	74
	Geology	4
	Mathematics	30
	Physics	44
Applied disciplines	Agronomy	16
	Applied mathematics	32
	Computer science	79
	Material science	99
	Medicine	8
	Metallography	60

Note: When there is perfect correlation, the correlation index with basic research is 100.

often inspired, induced and guided by knowledge innovation. In addition, during technological innovation, existing achievements of knowledge innovation are often referred to. Therefore, knowledge innovation is still at the upstream side of technological innovation, supporting it with knowledge.

2.2. *Development effect*

In addition to producing new knowledge, R&D activities can also increase the ability of innovation entities to absorb technologies. The production of new knowledge is closely related to the ability of innovation entities to absorb technologies. The generation of new knowledge allows the innovation knowledge base to grow and expand, thereby speeding up the learning of external knowledge and the identification of the utility of external knowledge in R&D. This enables quicker application of external knowledge in technological development. The ability of innovation entities to absorb technology is built upon the knowledge spillovers of competitors and the knowledge outside the industry. Only when the innovation entities are equipped with considerable internal technological capability can the innovation space be extended. How strong such capability is, however, must be reflected in the ability to make technological innovation. R&D capabilities lead to the increase in the innovation capacity and then expand the space for technological innovation. In this way, R&D has a developmental effect on technological innovation.

2.3. *Causal effect*

Through creative activities of R&D, original new technologies are born. Once R&D exerts development effect on technological innovation, new technologies will diffuse and will be imitated. In this case, R&D will generate a causal effect on technological innovation, which will then enter a continuous phase of expansion. In pursuit of potential economic benefits, relevant enterprises in the industry will increase R&D investment within the acceptable range of costs and risks. As R&D activities deepen, the proprietary knowledge of technological innovation continues to expand and the degree of knowledge coding continues to increase. Innovative conception continually converges on proprietary knowledge through coding, which includes diffusion and imitation, and finally forms the principle of innovation. However, when knowledge coding reaches a certain degree, the marginal effect of R&D on technological innovation will show a trend of decrease. In particular, when the cost of R&D activities is greater than that of converting new technologies, demanders of new technologies tend to obtain a more advanced technological platform through the diffusion and imitation of new technologies. This is the causal effect of R&D on technological innovation.

2.4. *Self-enhancing effect*

R&D activities continually exert stimulating, development and causal effects on technological innovation, boosting its development. The development of technological innovation is manifested by the increase in tangible output, namely, the expansion of direct business profit or market share, as well as by the growth in intangible output, that is, the acquisition and possession of more new knowledge and patents. The increase in tangible output will bring sufficient financial and material guarantee to the next phase of R&D activities while the growth in intangible output will provide it with more effective funding and technical support. This is a cyclical process with positive incentives, in which R&D exerts a self-enhancing effect on technological innovation.

3. The Progression Mechanism of Indigenous Innovation and Imitation

3.1. *Linkage between introduction and innovation*

A large number of technological capabilities are transferred across borders and there are a large number of advanced technologies in international transfer. On the one hand, this provides a rare opportunity for China to use external technological resources to develop new industries and promote industrial upgrading so as to achieve technological progress. On the other hand, the global allocation of technological factors has also brought unprecedented pressure on the development of China in many aspects. China's old strategy of "market for technology" has been terminated as it has resulted in the loss of a large amount of market share without obtaining necessary technologies. This fully demonstrates the cruelty of competition. In this book, we believe that the gap between China and developed countries can be narrowed and China can even catch up with them if the strategy of "linkage between technology and market" is adopted to realize the benign interaction and mutual promotion between introduction of technology and indigenous innovation. To be specific, On the one hand,

China may absorb the investment of MNCs in R&D internationalization at the expense of market and embrace their corresponding R&D institutions in China in exchange for advanced technologies from developed countries. On the other hand, on the basis of learning, digestion, absorption and imitation, technologies may be improved and innovated so as to create domestic and even international markets.

There is an essential difference between this strategy and the "market for technology" strategy. To clarify the relationship between the two, it is necessary to identify the reasons for R&D internationalization of MNCs and analyze why the "market for technology" strategy fails. A brief answer is as follows. On the one hand, the emergence of R&D internationalization investment is not a development model designed in advance by MNCs from developed countries to maximize their own interests. On the contrary, this trend appears because of the strategic adjustment that these MNCs are forced to make in the face of serious challenges on their profit model which they have used for years, challenges brought about by the rapid changes in global technological development and industrial division of labor as well as the continuous improvement in the investment environment of developing host countries. In other words, their transfer of technologies and manufacturing capabilities overseas is a reluctant and forced choice. There are two reasons. First, the development of modern technology, represented by information technology, has greatly accelerated technological upgrading and shortened product life cycle. For many products, competition already exists when they are still being developed, which forces MNCs to seek to synchronous use across the global or even transfer technologies in order to recover their huge investment in the shortest time possible. Second, the development of information technology has made synchronous R&D across the globe possible. Information technology allows R&D institutions of MNCs located in different countries to work on one project at the same time. Moreover, non-stop work around the clock in Europe, America and Asia greatly increase the speed and efficiency of R&D. On the other hand, the reasons for the poor performance of the "market for technology" strategy are manifold. First, relevant policies in China are imperfect, with problems like highlighting hardware and overlooking software. There is also a lack of information communication and coordination mechanisms, resulting in repeated introduction and vicious internal competition. Second, unless there are other constraints, MNCs' fundamental pursuit of profit decides that they will not transfer the most advanced technologies to China subjectively. Third, the asymmetrical strength of the two sides determines the effect of "market for technology" strategy. This is the most important reason because in the process of mutual game, the gap in strength between the two sides has a major impact on the final result. The above analysis, however, also indicates that as long as appropriate measures and methods are adopted and operations are rigorous, there is still possibility to acquire advanced technologies under the "market for technology" strategy.

The strategy of linkage between technology and market is put forward in a context that is distinct from that of the original "market for technology" strategy. With its accession to the World Trade Organization, China has begun to engage into the global market. The establishment of R&D institutions in China by MNCs is not simply about the Chinese market. On the contrary, attracting R&D investment from MNCs has become an open path adopted by China to make use of external technological resources and accumulate technological capabilities. Many enterprises in China still do not master the R&D capabilities for core technologies, yet they are still competitive and have the potential for growth. By virtue of their strengths in professional manufacturing, they do not have to undertake the high risk from huge investment in R&D, though they have no opportunities to reap huge profits from technologies. In particular, the foundation for division of labor is now changing. In the past, MNCs had both R&D and manufacturing capabilities. They only leave the products they are not willing to produce, including those they eliminated and those with low profit, in the developing countries, which is their passive choice. Nowadays, the basis for the pattern of division of labor is changing. For a good proportion of industries, the cost on local manufacturing prevents MNCs from participating in market competition and many advanced technologies and high-tech products are manufactured in China. This is actually the win– and mutually constrained cooperation between MNCs with R&D capabilities and core technologies and the host countries with the capacities for massive production and cost control.

To deliver the strategy of "linkage between technology and market", it should be noted that the linkage has two internal two-way cyclical processes. One is an inward loop in the following order: technology introduction, digestion and absorption, technological innovation, production and application, domestic diffusion, improvement in technical potential, new introduction of technology. The other is an outward loop in the follow order: technology introduction, digestion and absorption, technological innovation, production and application, export to increase foreign exchange earnings, and new introduction of technology. With mutual inter-action and infiltration, the two processes make up the cyclical dynamic process of technology introduction through dovetailing and embedding. How well the two loops work is decided by three links: introduction of technology, technological innovation and technology diffusion.

First, technology introduction is the basis for a virtuous cycle. The primary questions facing the introduction of technology are what kind of technology is to be introduced, whether the introduced technology is suitable, and how much benefit the technology introduced will bring. It should be noted that the introduction of technology must be aligned with the existing technological conditions so as to induce the "resonance effect" from the effect of introduction. Meanwhile, the resource condi-tions of China should also be considered to seek applicable technologies that match its resource conditions while attention is given to sustainable introduction. Lastly, it is necessary to correct the tendency of highlight-ing hardware and overlooking software in the introduction of technology and reduce dependence on foreign technologies. The introduction should cover both manufacturing technologies and composite technologies, including organizational, managerial and marketing techniques.

Second, technological innovation is the core of a virtuous cycle. The innovation activities after introduction are the core activities of the benign cycle of technology introduction. Technological innovation will emerge to meet new market demands, thereby bringing excess profits to techno-logical innovators and enhancing their competitiveness. Technological innovation requires that innovators must take market demand as the fun-damental starting point of technological innovation, acquire knowledge from the market, position new products, and carry out technological inno-vations. At the same time, the market should be developed through tech-nological innovation to change and create demand. A virtuous circle will be created through the start of the market.

Third, the diffusion of technological innovation is the true meaning and actual value of a virtuous circle. The true meaning and actual value of a series of technological innovation activities carried out after the intro-duction of technology lie in the proliferation of such innovation. The diffusion of technological innovation does not simply bring certain bene-fits to enterprises. Technological innovation can only create economies of scale through diffusion. As the subjects of technological innovation, enterprises should pay attention to the use of various technological diffu-sion resources. Technology diffusion among enterprises should be real-ized through technical cooperation as well as the purchase, sale and use of intermediate products and capital products produced by other enterprises (including competitors) so as to achieve economies of scale. The key to China's economic development and sustained economic growth lies in whether it has sufficient technical foundation. The core is the issue of technological innovation and technology transfer, that is, whether China can effectively integrate domestic and foreign advanced technological achievements into its own industrial system and take the lead in the world in terms of development of the original next-generation industrial system. If Chinese enterprises adhere to the strategy of "technology and market linkage" and take the market-oriented technological innovation as focus of this linkage process, they are likely to occupy a favorable position in the process of economic globalization and R&D internationalization.

3.2. *Progression of imitation and innovation*

In this section, mathematical models are built to study the progression of indigenous innovation strategy and imitative innovation strategy, so as to analyze the two as ways of corporate technological innovation, understand the relationship between the two, and affirm objectively at the macroscopic level that both are major means of technological innovation in China. In the context of the massive R&D internationalization of MNCs, how do the local enterprises of China choose a strategy of technological innovation that is in line with the national reality? This microscopic question will be answered in the following section that analyzes the transition between indigenous innovation and imitative innovation.

Indigenous innovation strategy and imitative innovation strategy have their own characteristics. The indigenous innovation strategy refers to the innovation behaviors of enterprises when they make breakthroughs in core technologies or concepts with their own efforts, based on which one should promote the follow-up parts of technological innovation by virtue of their own technological capabilities, realize the commercialization of technological achievements, and fulfill the expected goals for profit. Indigenous innovation is different from independent development. The former emphasizes the achievements of innovation. Its major sign is that the achievements of innovation enjoy independent intellectual property rights and the core technologies are owned by the enterprises that make such achievements. The latter emphasizes the form of innovation. It means making a certain achievement completely based on one's own strength. The achievements of indigenous innovation may include the citing of others' patent rights, but the overall system enjoys the characteristics of the achievements of indigenous innovation, i.e., the possession of independent intellectual property rights and the independent mastery of core technologies.

The imitative innovation strategy refers to the innovation behaviors of enterprises when they absorb and master the core technologies and technical secrets of pioneering innovators by means of introduction, purchase, reverse engineering and decoding, further develop and produce competitive products and participate in market competition. Imitative innovation does not mean completely copying others' technologies, as it also requires investment of certain R&D forces to further develop the technologies from pioneering innovators. Therefore, imitative innovation is not mere imitation, but a kind of progressive innovation behavior. Enterprises adopting imitative innovation do not act as the pioneering explorers or pioneering users of new technologies, but active learners of valuable new technologies.

Indigenous innovation and imitative innovation are the options of strategy for technological innovation. Correct selection of the strategy calls for analysis on the innovation capacity and innovation costs of enterprises. If every enterprise in the society is engaged in research on indigenous innovation, there will be so many repetitions that the output to input ratio of the research process across the market is too low. This then means that the R&D costs of the whole market are high, yet technology progress may not necessarily be fast. MNCs are always at the forefront of indigenous innovation based on technology leadership strategies. Meanwhile, their R&D achievements generate technology spillover effects that provide competitors with opportunities for learning and imitation. The competitors can also improve their level of technology with the strategy of imitative innovation and form the evolutionary pattern that social indigenous innovation and imitative innovation progress at the same time.

Suppose that there are a number of enterprises in the market that produce similar products competing with each other. They are trying to capture greater market share and gain excess profits through technological innovation. The technology leaders represented by MNCs choose the strategy of indigenous innovation while the technology followers represented by local enterprises follow the strategy of imitative innovation. According to the hypothesis and deduction of models for R&D knowledge spillover effect by Hou *et al.* (2001), the production function is

$$Y = AL^{1-\alpha-\beta}H^{\beta}\sum\nolimits_{j=1}^{N}(X_j)^{\alpha}, 0 < \alpha < 1, 0 < \beta < 1;$$

the quantity of the *j*th kind of intermediate products used is

$$X_j = LA^{1/(1-\alpha)}\alpha^{2/(1-\alpha)}h^{\beta/(1-\alpha)};$$

the per capita output function is

$$y = A^{1/(1-\alpha)}\alpha^{2\alpha/(1-\alpha)}h^{\beta/(1-\alpha)}N;$$

the wage rate is

$$w = (1-\alpha-\beta)y;$$

the interest rate is

$$r = (L/u)\left(\frac{1-\alpha}{\alpha}\right)A^{1/(1-\alpha)}\alpha^{2/(1-\alpha)}h^{\beta/(1-\alpha)};$$

the technical growth rate is

$$g = \frac{1}{\sigma}\left[\frac{L}{u}\frac{1-\alpha}{\alpha}A^{1/(1-\alpha)}\alpha^{2/(1-\alpha)}h^{\beta/(1-\alpha)} - \rho\right].$$

In the above equations, Y represents the quantity of the final products, A represents the parameters of labor productivity that reflects the government's policies, such as technological level, protection for property rights and taxation, L is the non-technical labor input, h is the level of human capital stock, X_j is the quantity of the *j*th intermediate products used, and N is the number of types of intermediate products. The increase in N indicates R&D knowledge innovation or technological progress. $H = H/L$ is the density of human capital, $y = Y/L$ is the per capita output, w is the wage rate, r is the interest rate, u is the R&D cost, g is the technological growth rate, σ is the relative risk factor, and ρ is the discount rate.

Firstly, we analyze the technology spillover effects of MNCs. MNCs enjoy the technological advantages derived from indigenous innovation (relevant parameters are marked with subscript Z). Local enterprises only achieve imitative innovation based on the technologies of the MNCs developed through indigenous innovation (relevant parameters are marked with subscript M). Then, obviously costs of indigenous innovation are greater than those of imitative innovation, i.e., $u_Z > u_M$. The sets of indigenous innovation products contain those of imitative innovation products, i.e., $N_Z > N_M$. If indigenous innovation and imitative innovation are in the same stable state, i.e., $g_Z = g_M$, then the costs of imitative innovation in a stable stage u_M^* can be figured out (the corresponding costs of indigenous innovation of MNCs is u_Z). Hence,

$$u_M^* = u_Z (L_M / L_Z)(A_M / A_Z)^{1/(1-\alpha)} (h_M / h_Z)^{\beta/(1-\alpha)}.$$

When $u_M < u_M^*//u_Z$, there exists $g_M > g_Z$. So, N_M increases and g_M decreases until $u_M = u_M^*$, $N_M = N_Z$, and $g_M = g_Z$. In other words, if the costs of imitative innovation are lower, i.e., the technological gap between the two sides is relatively large, then the technological growth rate of imitative innovation is faster than that of indigenous innovation. The types of products produced by technologies from imitative innovation are growing faster than those by technologies from indigenous innovation, but the acceleration of technological growth obtained by imitative innovation declines until the technological growth rate of imitative innovation is equal to that of indigenous innovation. After this, MNCs continue to engage in indigenous innovation and local enterprises immediately imitate them. This is a process of imitative innovation catching up with indigenous innovation, as shown in Figure 2.

As can be seen from Figure 2, although the indigenous innovation of MNCs is ensuring technological growth at a relatively constant speed and the technological gap between local enterprises and MNCs is large, the imitative innovation of local enterprises is effective and the speed of technological progress increases until it is close to that of indigenous innovation. Then the technological growth rate of the two is similar. In this case,

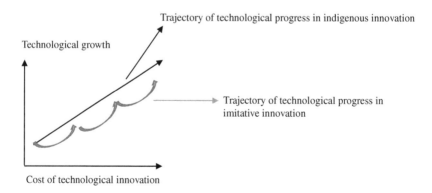

Figure 2.   Progression Mechanisms of Indigenous Innovation and Imitative Innovation.

there is a trend of a relatively steady state between indigenous innovation and imitative innovation. In other words, the pattern of converging to a stable state emerges. At the same time, the technologies that local enterprises can imitate are reduced, the relative speed of technological progress declines and the technological gap between MNCs and local enterprises widens again. Technology spillovers of MNCs are generated, leading to a new round of technological progress of local enterprises. In this way, a kind of spiral progression of indigenous innovation and imitative innovation appears. Each round of progression drives technological progress and promotes economic growth.

3.3. *Transformation of imitation and innovation*

Many technology strategists have analyzed the advantages and disadvantages when enterprises of different sizes adopt indigenous innovation, imitative innovation and cooperative innovation and discussed the positioning and selection of strategies of technological innovation at different stages of the life cycle of enterprises.[1] In fact, for a single enterprise, indigenous innovation and imitative innovation are not an either-or option. With the changes in its own technological capabilities and market strategies, the strategies of technological innovation can be transformed from one to the other. From a microscopic point of view, a single enterprise is open to the two options, indigenous innovation and imitative innovation at any time. It will definitely choose the best strategy by analyzing their costs and benefits. Besides, since the two decision-making models can be transformed into each other, the choice of strategy is dynamic and changing, and will automatically reach a stable structure. The mathematical models are established in this way.

Suppose that there are several enterprises in the market that produce similar products competing with each other. Each enterprise has two options of strategy: indigenous innovation and imitative innovation. The probability of enterprise making investment in indigenous innovation in the T period is p_t $(0 < p_t < 1)$, with total costs C_Z and benefits R_Z. Then the income of indigenous innovation is expressed as follows:

$$R_Z(p_t) = f_Z(K, p_t, T_Z) - C_Z \text{ and } f_Z(K, p_t, T_Z) = \frac{KT_Z}{1 + p_t},$$

---

[1]Central Innovation Network. http://www.hb.xinhuanet.com/zhuanti/2004-06/01/ content_2226457.htm.2004-06-01.

where $f_Z (K, p_t, T_Z)$ is the benefit function of indigenous innovation determined by the business scale K of the enterprise, the investment probability p_t in the period t, and the technological factor T_Z.

According to the literature, when the business scale of the enterprise is larger and the technological factor is higher, the economic benefits will be higher (Yi *et al.*, 2005). When the probability of investment in indigenous innovation is higher, the success of indigenous innovation will be greater, which will intensify the competition among enterprises and reduce their economic benefits.

According to the same assumptions mentioned above, the probability of investment in imitative innovation is $1 - p_t$, with total costs C_M and benefits R_M. Then the benefits of imitative innovation are expressed as

$$R_M (p_t) = f_M (K, p_t, T_M) - C_m.$$

Similarly, $f_M (K, p_t, T_M) = \frac{KT_M}{1+p_t}$, $f_M (K, p_t, T_M)$ is the benefit function of indigenous innovation determined by the business scale K of the enterprise, the probability of investment p_t in period t and the technological factor T_M. When the business scale of the enterprise is larger, the economic benefits it obtained are higher, the probability of investment in indigenous innovation is higher and the possibility of imitative innovation and its economic benefits are lower.

Next, suppose that each enterprise can gradually adjust its innovation strategy in the process of technological innovation, i.e., when the benefits from adopting the strategy of indigenous innovation is lower than the average social benefits, the enterprise can withdraw from indigenous innovation and go for the strategy of imitative innovation. Alternatively, when it uses the strategy of imitative innovation and discovers that the indigenous innovation may bring greater profits, it can also go for it. Therefore, the benefits of individual enterprises adopting indigenous innovation or imitative innovation are

$$R(p_t) = p_t R_Z (p_t) + (1 - p_t) R_M (p_t).$$

According to the literature, the changes in the probability of enterprises investing in technological innovation in each period can be described by the following discrete dynamic system (Hofbauer, 1998), i.e., if the benefits from an enterprise's investment in indigenous innovation are greater than those obtained by the average individual in the society, then the probability of the enterprise making investment in indigenous innovation in the next phase will increase. Otherwise, such probability goes down. This can be expressed as

$$p_{t+1} = p_t \frac{R_Z(p_t)}{R(p_t)}.$$

Subtract p_t from both sides and put total benefits into the equation. It then becomes

$$p_{t+1} = p_t + \frac{p_t(1-p_t)(R_z - R_m)}{p_t(R_Z - R_m) + R_m}.$$

Let $p_{t+1} = p_t$, then there must be $R_Z = R_M$. Hence,

$$f_Z(K, p_t, T_Z) - f_M(K, p_t, T_M) = C_Z - C_M.$$

Put f_Z and f_M into the above equation and $\frac{T_Z - T_M}{C_Z - C_M} = \frac{1+p_t}{K}$. This equation has the following meanings:

First, the difference between the technological benefit factor brought about by technological progress from indigenous innovation and the technological benefit factor of imitative innovation has a constant proportional relationship with the difference between the costs of indigenous innovation and those of imitative innovation. Such proportional relationship is related to business scale and the probability of indigenous innovation. It is positively related to investment in indigenous innovation and negatively related to business scale.

Second, the probability of investment in indigenous innovation is a balanced system. A balance is found between indigenous innovation and imitative innovation through the dynamic adjustment by enterprises. When the benefits of indigenous innovation are higher, enterprises will increase investment in indigenous innovation, but risks will also increase. On the contrary, when the profits of imitative innovation are high, enterprises will also be tempted to avoid the risks of indigenous innovation.

Third, the probability of investment in indigenous innovation is converging. It does not matter which of the two kinds of investment has a higher probability, it will eventually reach equilibrium. The rate of convergence depends on the difference in benefits between indigenous innovation and imitative innovation. If the difference is smaller, the probability of investment in indigenous innovation converges faster. Otherwise, the probability is more likely to oscillate.

Yi (2005) shows that under certain conditions, the proportion of enterprises making investment in indigenous innovation in the market and the market structure will evolve steadily. He uses numerical simulation to analyze how market is stabilizing after long-term dynamic adjustment under different conditions. According to the previous analysis and conclusions, the rational decision-making process of indigenous innovation and imitative innovation of a single enterprise is as follows. If the costs of indigenous innovation of an enterprise increase, the probability of its investing in indigenous innovation will decrease, which will lead to the reduction in achievements of indigenous innovation. Meanwhile, the costs of imitative innovation are relatively smaller in this stage and the achievements of such innovation increase. However, in the next round of imitative innovation, due to the decline of achievements of indigenous innovation in the previous round, the opportunities for imitative innovation diminish, which skews the enterprise toward indigenous innovation in another round of decision-making. The benefits of the last decision for technological innovation of an enterprise will to a large extent affect the decision-making in the next round, yet the impact is less than 100%. Hence, the system is a negative feedback system, which is the fundamental cause of convergence.

4. Test on the Influence of R&D Internationalization on Indigenous Innovation

4.1. *Construction of model*

We draw on the output model to construct the following model for the relationship between the level of R&D internationalization and regional indigenous innovation capability:

$$Y = \alpha I^{\beta} \mu, \tag{1}$$

where I and Y are the comprehensive indicators of R&D internationalization and indigenous innovation, both of which are derived from the actual measurement above. They are standardized variables without units.

We use the panel data analysis method. In order to compare the empirical results, we take the logarithm of both sides of Eq. (1) and obtain the logarithmic model (2). In this way, the coefficients before the explanatory variables represent the concepts of elasticity.

$$\ln Y_{it} = \alpha + \beta \ln I_{it} + \mu, \tag{2}$$

where i is the province/autonomous region/municipality and t is the year.

4.2. *Empirical test*

We first examine interprovincial samples from China and analyze the overall effect of R&D internationalization on the indigenous innovation. Since there are huge differences in the development of indicators for R&D internationalization and indigenous innovation among eastern, central and western regions in China, the relationship between the two and their mechanisms of action should also be different. To test such differences, all the samples are divided into two groups, namely, based on eastern and central/western regions,[2] for the analysis of the influence of R&D internationalization on indigenous innovation. We use a mixed model for analysis. The results of the analysis are shown in Table 3.

According to the regression results, all variables passed the significance test at the level of 5% and the adjusted R^2 is higher, which indicates that the model works well and can be used to explain the reality.

Table 3.   Regression Results of Mixed Effects Model (2000–2005)

	General	**Eastern region**	**Central/Western regions**
ln I	0.038	0.078	0.010
C	1.783	1.423	1.860
Adjusted R^2	0.783	0.751	0.710

---

[2]The eastern region includes 11 provinces and municipalities, including Beijing, Tianjin, Hebei, Liaoning, Shanghai, Jiangsu, Zhejiang, Fujian, Shandong, Guangdong and Hainan. The central region includes eight provinces, including Shanxi, Jilin, Heilongjiang, Anhui, Jiangxi, Henan, Hubei, and Hunan. The western region includes 12 provinces, municipalities and autonomous regions, including Chongqing, Sichuan, Guizhou, Yunnan, Tibet, Shaanxi, Gansu, Qinghai, Ningxia, Xinjiang, Guangxi and Inner Mongolia. Since Tibet's data are difficult to obtain, this book uses a total of 30 provinces for analysis.

In other words, R&D internationalization has a significant positive effect on indigenous innovation on the whole as well as in the eastern, central and western regions separately. Among them, the eastern region is more obvious. When R&D internationalization in a province of the region increases, its indigenous innovation capacity is greatly boosted.

Specifically, for every 1% of increase in the level of R&D internationalization, the indigenous innovation capacity goes up by 0.038% on the whole, by 0.078% for the provinces in the eastern region and by 0.010% for the provinces in the central/western regions.

4.3. Factors behind the influence of R&D internationalization on indigenous innovation

According to previous studies, the regional market size, scale of foreign investment, degree of protection for intellectual property rights, R&D resources and infrastructure construction are external factors that have an important impact on indigenous innovation. Therefore, from those five aspects, we examine the factors influencing how R&D internationalization affects indigenous innovation as follows:

(1) **Regional market size:** We use the annual average of GDP of the provinces from 2000 to 2008 (unit: ¥100 million) to measure their market size. The 30 provinces are divided into two groups based on the annual average of GDP.

(2) **Scale of foreign investment:** We use the annual average of FDI stock of the provinces from 2000 to 2008 (unit: $10,000) to measure their scale of foreign investment. With the actually utilized FDI flow indicator of the regions, we adopt perpetual inventory method to measure the FDI stock of the regions. For the stock in base period, the investment flow of the base year is divided by 10% for estimation. Considering the availability of data, we draw on the above method, with 1985 as the base year,[3] calculate the stock of 1985 and use it as the base stock. We also assume a depreciation rate of 9.6% and figure out the FDI stock from 2000 to 2008 based on the perpetual inventory method. The 30 provinces are divided into two groups based on the annual average of FDI.

---

[3]Data on Beijing, Inner Mongolia, Hubei, Qinghai, and Sichuan are not complete. Therefore, the calculation is made based on those years with data using the same method.

(3) **Degree of protection for intellectual property rights:** We use the annual average of patents owned by every 10,000 people of the provinces from 2000 to 2008 (unit: piece per 10,000 people) to measure their degree of protection for intellectual property rights. The 30 provinces are divided into two groups based on the annual average of number of patents owned by every 10,000 per year.

(4) **R&D resources:** We use the annual average of the R&D funds of the society at large of the provinces from 2000 to 2008 (unit: ¥10,000) to measure their abundance of R&D resources. The 30 provinces are divided into two groups based on the annual average of R&D funds of the society at large.

(5) **Infrastructure construction:** We use the annual average of the local public expenditure on infrastructure construction of the provinces from 2000 to 2008 (unit: ¥10,000) to measure their infrastructure construction. The 30 provinces are divided into two groups based on the annual average of local public expenditure on infrastructure construction.

The GDP data come from the *China Statistical Yearbook* from 2001 to 2009. The R&D fund of the society at large and the number of patents owned by every 10,000 people are derived from the *China Statistical Yearbook on Science and Technology* from 2001 to 2009. The FDI stock is calculated based on the data from the China Economic Information Network as well as the statistical yearbooks, almanacs of economy and statistical communiqués of the provinces and municipalities, using the perpetual inventory method.

Each group is analyzed with mixed effects model. The regression results are shown in Table 4.

According to the regression results, all variables passed the significance test at the level of 5% and the adjusted R^2 is higher, which indicates that the model works well and can be used to explain the reality.

In the case that all the factors have significant influence in the two groups, we conduct a Chow test for parameter stability on the factors and analyze whether there is structural stability between each two groups of samples, i.e., whether there is significant difference in the regression coefficients between each two groups of samples. As the results of Chow test reveal, there are significant differences in market size, scale of foreign investment, protection for intellectual property rights and infrastructure construction. There is no significant difference in R&D resources, i.e., R&D resources have no obvious impact on the promoting effect of R&D internationalization on indigenous innovation.

Table 4. Grouped Regression Results of Mixed Effects Model of Each Factor (2000– 2008)

	Market size		Scale of foreign investment		Degree of protection for intellectual property rights		R&D resources		Level of infrastructure construction	
	High	Low	High	Low	High	Low	High	Low	High	Low
ln I	0.145	0.006	0.041	0.008	0.087	0.006	0.038	0.046	0.129	0.013
C	1.588	1.867	1.927	1.862	1.794	2.862	21.904	1.825	1.645	1.888
Adjusted R^2	0.992	0.998	0.999	0.999	0.999	0.999	0.997	0.997	0.991	0.997

According to the regression results, in regions with a larger market size, more foreign investment, higher degree of protection for intellectual property rights and sound infrastructure, the level of R&D internationalization plays a significant role in improving indigenous innovation capacity, and vice versa. This shows that market size, foreign investment, protection for intellectual property rights and local infrastructure all facilitate the positive influence of the level of R&D internationalization on indigenous innovation capacity, which is consistent with theoretical analysis and common sense. However, R&D resources have no obvious influence on such positive influence. Even viewed from the coefficient of explanatory variables, the influence of R&D internationalization on indigenous innovation in regions with abundant R&D resources is weaker than that in the regions with less R&D resources. We believe that this is probably because R&D resources as an influencing factor is different from other factors and it has a stronger cognate relationship with the independent variable, i.e., the level of R&D internationalization. Regions with rich R&D resources also have higher indigenous innovation capacity. Foreign R&D capital, together with domestic R&D resources, will exert an impact on the regional indigenous innovation capacity. This kind of positive effect is diminishing marginal effect. That is to say, if a region has richer R&D resources, the increase in foreign R&D capital has a smaller promoting effect on the improvement of regional indigenous innovation capacity. Based on this, a very interesting point can be raised that R&D has a significant positive impact on indigenous innovation. Yet, such an impact is not unlimited; it is constrained by other factors.

Chapter 5

The Inhibitory Effect of R&D Internationalization on Indigenous Innovation in China

1. Inhibitory Effect from One-Way Talent Flow

1.1. Channels of one-way flow of talent

As China continues to implement and deepen its reform and opening-up, MNCs are gradually shifting their headquarters and R&D institutions to China. This reflects the importance that MNCs are attaching to the Chinese market and their hope to build a complete value chain from R&D through production to sales in China so as to enhance their competitiveness.

Generally speaking, the impact of the development of foreign R&D institutions on host countries tends to be complex, with both positive and negative effects. In terms of negative impact, the R&D of MNCs in China will bring unfavorable results such as one-way flow of talent and crowding-out effect of competition. As known to all, human capital, especially high-quality professionals, is a major booster of the international competitiveness of an enterprise as well as a nation. One of the purposes for MNCs to invest in R&D in China is to use China's high-quality yet cheap talent. Hence, fierce competition between the talent in MNCs and that in Chinese enterprises is inevitable. China possesses not only rich and premium technological human resources with lower wage costs but also a large number of high-quality scientists and engineers who can carry out

R&D at all levels from basic research to development of new products. Every year, more than 4 million college students enter the workplace in China, which is more than four times of that in the United States. The number of software professionals in China is twice more than that in India. Therefore, in the recent years, MNCs have started to pay increasing attention to the R&D force in China and approach them in a different light.

It has been a tradition in China to highlight the significance of education. Chinese people have a strong desire to lift themselves out of poverty and backwardness through education and hard work. On the whole, China has a strong awareness about fostering human resources. Besides, Chinese workers are more diligent and more willing to work than those in the developed countries. On the other hand, because of the adequacy of resource endowments, the majority of Chinese workers, including those engaged in science and technology, require far less payment than those in the developed countries.

In the recent years, China has attached more importance to English education to make it easier for Chinese scientists and technicians to communicate with their foreign counterparts. Many of the technological professionals who have completed their studies abroad and returned to China also enter the job market. There exists a prominent trend of English becoming a working language. China has an increasing number of professionals necessary for the development of various kinds of new technologies and products. The utilization of this advantage in human resources makes it cost effective for MNCs to invest in China. The availability of a large amount of talent has paved the way for MNCs to set up R&D institutions in China.

In this context, China's R&D institutions and enterprises face severe challenges, and the possible outcome is that the talent of domestic research institutions "flows out". The huge market demand coupled with plenty of high-quality human resources have undoubtedly made China the ideal strategic base for MNCs to set up their R&D institutions. In the software industry, for example, the cost of software talent in the United States is nine times that in China and the cost of software talent in India is twice that in China. A Chinese professor or a researcher equivalent to a professor in a developed country has an annual salary of only tens of thousands of yuan. The average wage of Chinese R&D professionals is much lower than that of their counterparts in the developed countries. The one-way flow of talent will make it increasingly difficult to keep the domestic technologies confidential. In R&D centers funded by MNCs, R&D professionals from mainland China account for 95% on average. In the R&D centers of some well-known MNCs, such as Beijing Samsung Communication Technology Research Co., Ltd., Microsoft China Research and Development Center, Motorola China Research Institute, and Bell Labs of Lucent Technologies (China) Co., Ltd., almost all the R&D professionals come from mainland China, with an average proportion exceeding 99%.

1.2. *Consequence of one-way flow of talent*

For R&D investment in China, MNCs have a significant demand for local high-tech talent, especially the in-service researchers of Chinese universities, research institutions and enterprises, which are their first choice. As a result, the one-way flow of talent from Chinese innovation-based organizations to MNCs and the loss of talent of such organizations in the short term become inevitable. Moreover, the elite among R&D professionals who flow to MNCs may even be selected and sent to work in their home countries, resulting in cross-border flow of talent. The reduction of technological strength will have a huge impact on China's R&D activities and increase the technological dependence of domestic enterprises. Senior technological and management professionals tend to pursue high salary and high efficiency. In addition, they also require a sound work environment and atmosphere to fulfill their value. The R&D institutions established by MNCs in China usually provide jobs that are more challenging, with better payment and working conditions, than domestic enterprises, which makes it easier for those senior technological and management professionals to give full play to their competence and improve their living conditions. As a result, most of them have flowed from domestic enterprises and research institutions to MNCs. According to statistics, since the 1990s, when MNCs in the IT industry, such as IBM, HP, and Microsoft, crowded into the Chinese market, Zhongguancun has lost half of its IT professionals, most of whom have gone to these MNCs.

Technology is a kind of intangible asset that is accumulative and investable, as evidenced in the allocation of its carrier — human resources. The proportion of high-quality professionals in China is not high. They are flowing domestically, yet one way from domestic enterprises and research institutions to MNCs because the latter offer higher payment, which is strongly appealing to the Chinese research professionals.

The one-way flow of talent brings two serious effects. First, local enterprises lack direct motivation for indigenous innovation due to the lack of talent. The core of competition among modern enterprises is innovation, whose foundation is talent. It is self-evident that indigenous innovation is an empty slogan without talent for innovation. Second, and more importantly, the talent used to be carriers of technological innovation of domestic enterprises and research institutions. Personally, they have very rich experience in technological development; yet, they also command a large amount of technical information. When they serve with R&D institutions of MNCs, with the cultural mentality that one good turn deserves another, they tend to be willing to give full play to their strength. In this way, the advanced technologies in some areas of China are acquired by MNCs at low costs, which is reverse technology spillover toward MNCs.

The one-way flow of talent caused by the R&D of MNCs in China does not simply have a negative impact on Chinese domestic enterprises and research institutions. It is even likely to severely weaken China's innovation capacity and thus international competitiveness, bringing unfavorable impact on the long-term development of the economy. First, for domestic companies and research institutions, due to the one-way flow of talent, technical experience that the professionals have accumulated over the years is obtained by the R&D institutions of MNCs at a small price through R&D. Compared with most domestic enterprises, MNCs have stronger international competitiveness. One-way flow of talent may further widen this gap. This vicious circle may lead to a decline in the competitiveness of Chinese domestic companies, which may lose not only their international competitiveness but also their domestic market. In addition, China's protection system for intellectual property rights is still not comprehensive, and thus, there is no effective protection measure. In this context, the one-way flow of talent will cause serious losses to China and pose a certain threat to China's technological security. Second, the one-way flow of talent may also have an adverse impact on China's indigenous innovation. One-way flow of talent, the subject of innovation, causes massive talent outflow from Chinese enterprises and research institutions, which is very unfavorable for China's indigenous innovation.

1.3. *Countermeasures for curbing one-way flow of talent*

It is an urgent task to enhance the indigenous innovation capacity and R&D level of Chinese enterprises so as to curb the one-way flow of talent.

We believe that, on the one hand, it is necessary to reform the imperfect human capital development and management system, limit the influence of MNCs to the technical level as much as possible and adjust our talent policies to attract outstanding talent at home and abroad, and curb the trend of talent outflow. On the other hand, we must work hard to create a social atmosphere of respecting knowledge and talent. We must endorse the view that technology is the primary productive force, conscientiously implement the strategy of rejuvenating the country through science and education, adjust talent policies and effectively solve problems concerning the social status, working conditions and economic treatment of senior technological professionals so as to create a good working and living environment for them and encourage them to engage in technological invention and basic research. Hence, we should abolish some outdated or unreasonable policies and other backward views of employment as well as help Chinese enterprises develop a correct concept of R&D talent.

In addition, the external environment for innovation must be optimized. Driven by their career pursuit, scientific research professionals tend to seek a scientific research environment conducive to innovation. With abundant capital, MNCs can provide a superior scientific research environment to effectively unlock the talent of professionals. This is also an important reason why MNCs are attractive to scientific research professionals in China. For the professionals, what MNCs can provide is conducive to their development. Hence, from another perspective, MNCs reserve professionals for China. Therefore, optimizing the environment for scientific research and entrepreneurship is an important means to attract the talent back. China may establish a software and hardware environment conducive to entrepreneurship, such as venture capital and business incubator, to attract the talent to start businesses back in China and bring back advanced technologies, which is similar to the technology spillovers of MNCs to China.

The globalization of R&D is not only a major part of the global strategy of MNCs but also an effective way for them to carry out knowledge innovation on a global scale. The extent to which a country can access and utilize these types of R&D resources is increasingly dependent on the country's attractiveness to the owners of these resources. Therefore, countries around the world highlight and actively respond to the improvement of the national innovation system, investment environment and business environment as well as the development of higher education system. How to attract R&D investment from MNCs is an item on their agenda.

For example, Singapore has taken measures such as double tax cuts, investment subsidies, and accelerated equipment depreciation for important R&D activities that MNCs carried out after January 1, 1981. China should pay full attention to the trend of R&D globalization of MNCs, formulate new strategies for attracting investment and technologies, and actively seek the best partners for technological cooperation development. Meanwhile, the investment of MNCs in China should be incorporated into China's industrial integration system, so that their localized R&D can be carried out in depth and can form a virtuous cycle with China's technological progress. In this way, we can get rid of the technological catch-up traps and use second-mover advantage to promote technological progress in China and ultimately realize the leap-forward development of China's productive forces.

2. Analysis of Crowding-out Effect of Competition

2.1. *Forms of crowding-out effect of competition*

The crowding-out effect of competition from the R&D of the MNCs mainly includes technological crowding-out effect and market crowding-out effect. These two kinds of crowding-out effect interact and intertwine with each other, which imposes enormous external pressure on the domestic enterprises.

First, technological crowding-out effect means that Chinese enterprises are crowded out of the market by foreign enterprises because of the great advantages of MNCs in R&D. The R&D strategies of MNCs in China have confronted Chinese enterprises with greater competition. After establishing R&D institutions in China, MNCs have increased their R&D focused on the Chinese market and greatly enhanced their competitiveness in the Chinese market. The research results of their R&D institutions are mostly used by their subsidiaries in China, which helps them to form technological advantages and occupy a favorable position in the process of taking up the Chinese market. The main goal of MNCs setting up R&D institutions in China is to occupy the vast Chinese market for profits. To achieve this goal, they will transfer some technologies, but not the key ones concerning their core competitiveness. This will inevitably lead to fiercer competition for Chinese enterprises, resulting in the decline of China's industrial competitiveness and the gradual loss of second-mover advantages. It will also gradually enhance the competitiveness of MNCs, further widen the gap in industrial competitiveness between China and developed countries and force some backward enterprises to withdraw from the market. As a result, more Chinese enterprises will go bankrupt and more people will lose their jobs, which will add to social instability. According to a report on *Economic Daily* published on May 29, 2004, in the face of the patent challenges from MNCs, some well-known Chinese pharmaceutical companies decided to gradually leave the pharmaceutical industry. There is no doubt that the entry of the R&D institutions of MNCs into China will have a series of impacts on China's R&D.

Second, the market-crowding effect is essentially a follow-up effect to the technological crowding-out effect. Generally speaking, technological crowding out can cause decline in competitiveness and then the loss of market. This kind of crowding-out effect and the technological crowding-out effect are not much different from the first case. But in fact, the market crowding-out effect is not limited to the above-mentioned situation. Since MNCs are generally large internationally renowned companies, they have considerable advantages of scale in terms of output. This kind of scale effect can reduce their production costs. Moreover, MNCs can also take advantage of the low-cost effect of human resources in China to bring their production costs to far lower than similar companies. MNCs can occupy markets of host countries at a lower cost and push the domestic enterprises out of the market.

In short, the exclusive scale advantages and technological advantages of MNCs as well as the unique advantages of low-cost human resources in China pose huge challenges to the domestic enterprises. Compared with MNCs, domestic companies have no advantage in this regard. Once they lose market competitiveness, they will lose the impetus and financial resources for R&D. Such vicious cycle may not only cost domestic enterprises the international market but also make it increasingly difficult for them to maintain their competitiveness in the domestic market.

2.2. Negative impact of crowding-out effect of competition

MNCs have a large business scale, with operations in many countries around the world. Since the direct or indirect motivation of the MNCs is to achieve their goal of maximizing profits, they have little or even no concern about the industrial competitiveness of host countries being undermined or crowded out due to the entry of their R&D. Developing countries are particularly concerned about the problem that domestic enterprises may be crowded out by foreign subsidiaries due to the entry of foreign capital. Being crowded out by R&D may happen in two areas. First, in the product market, MNCs enter the markets of host countries and have an adverse impact on the learning and growth of local enterprises in the competitive industries, thus crowding out local enterprises. Second, in the financial market and other factor markets, MNCs reduce the funds or other factors available to local enterprises, or increase their cost on acquiring funds and factors, or do both, to crowd out local enterprises.

Being crowded out greatly undermines the core technologies and indigenous innovation capacity of local enterprises, which are precisely the two key factors for enterprises to participate in international competition. At present, many Chinese enterprises are at a disadvantageous position in international competition due to the lack of core technologies and indigenous innovation capacity. According to relevant surveys, color TVs, computers, DVDs and mobile phones are the main products in the electronic information industry in China, but most of the intellectual property rights of core technologies for these four types of products are not in the hands of Chinese enterprises. The lack of core technologies has resulted in low economic efficiency and outflow of wealth in China. For example, the wholesale price of an inferior DVD produced in China is $40, of which patent fee takes up more than 50%. The *2005 Report of Transnational Corporations in China* reports that, while foreign capital brought economic prosperity, China's gross national income (GNI) did not achieve the same rapid development as GDP. Before 1992, China's GNI was basically the same with GDP, or sometimes slightly higher. However, from 1993 to 2003, GNI in China became smaller than GDP, and the gap has been widening every year. This gap shows that part of the value created in China does not become the income of Chinese residents, but flows to other countries and becomes the wealth of their residents.

The technological crowding-out effect also has a negative impact on the development of some industries. Chinese enterprises lack indigenous innovation capacity. Yet, it is impossible for MNCs to transfer their core technologies to them. Rather, MNCs often squeeze the living space of Chinese enterprises from the source of the industries by means of protecting intellectual property rights and accelerate their pace in seizing the Chinese market.

At present, the technological and market crowding-out effects of MNCs on domestic enterprises have already emerged.

First, since the power of introducing new products is controlled by the foreign parties, it is impossible for the joint venture to carry out innovative activities that violate the interests of the parent company of the foreign parties.

Second, when the product rights are dominated by the foreign parties, they hardly allow an active R&D organization to exist in the joint venture, which has led to the loss of R&D power when Chinese enterprises form a joint venture with foreign parties.

Last, due to the lack of indigenous innovation capacity and competitive edge in cost, some industries in China are gradually losing their high-end markets.

2.3. *Countermeasures for inhibiting crowding-out effect of competition*

It should be noted that the first problem of the crowding-out effect of the R&D activities of MNCs on host countries suggests that we need to properly protect domestic enterprises but not in the sense of traditional trade protectionism. The elimination of trade protection will enable domestic consumers to obtain benefits, such as cheaper inputs and diversified products, but some of the domestic production and employment may suffer losses accordingly. In contrast, even if the protection against R&D competition is removed, the interests of domestic consumers will not increase due to the decline in domestic production, but the development of domestic entrepreneurship will be limited, especially in the frontier production areas. As a result, the association between local enterprises and R&D is weakened and technological deepening is inhibited.

However, it is important to distinguish the two situations of crowding out. One is that foreign subsidiaries crowd out domestic enterprises that have potential competitiveness. The other is that foreign enterprises crowd out inefficient local enterprises that lack competitiveness. Obviously, the second case of crowding out will help improve the overall competitiveness of domestic enterprises in China, making the economy more dynamic and flexible.

The second kind of crowding out reflects the unbalanced competitive conditions that arise in the domestic market due to local market segmentation. With reputation and scale, MNCs enjoy the special advantage in acquiring factors, such as capital and convergence of talent. The R&D

activities of MNCs will therefore increase the entry costs of local enterprises or enable them to get the best factor input at a reasonable price.

These two kinds of crowding out have both evoked policy concerns. Most governments want to promote the development of local enterprises, especially in the industrial activities that are technologically complex and dynamic. Many governments believe that the benefits from deepening the capabilities of local enterprises are greater than those from acquiring the same technologies from MNCs because local enterprises can strengthen their ties with local suppliers and increase interactions with local institutions. Besides, in terms of innovation activities, the knowledge developed internally will not be "exported" to the parent companies for application abroad. Developing countries that have developed advanced local technological capabilities almost universally do not limit FDI or restrict FDI from entering certain areas.

Facts have shown that having a strong local technological base is extremely important for local enterprises to develop competitiveness and attract high-tech R&D. After all, MNCs need local talent, technological infrastructure and local technical knowledge for their R&D, and they are integrated to make innovation or facilitate the creation and improvement of innovation. Therefore, the strategy that developing countries should adopt is to be self-reliant and exchange technology for technology. In practice, it is important to distinguish crowding out from legal competition. Policy makers should make such distinction and develop sound policies.

3. Analysis of Defensive Follow-up Strategy

3.1. *Forms of defensive follow-up strategy*

One of the main manifestations of defensive follow-up strategy is the defensive innovation strategy, which means that an enterprise chooses to defend their technological innovation and product innovation in a certain period of time and waits for opportunities to take actions. Proactive defensive strategy is adopted to avoid the unpredictable market risks caused by taking the lead. The defensive strategy is based on the principle of not striving for the first. It emphasizes that we have what others also have; if theirs are new, ours must be good; we do not need the newest but the best that are reliable. It is a kind of strategy of technological innovation aimed at safe operation without taking major risks. The main difference between a defensive innovation strategy and an offensive innovation strategy is the nature and time of innovation. Neither of them is willing to be the first innovator or a laggard. They expect to learn from the experience of the early adopters of innovation, follow them closely, and seize the favorable opportunity when the entrepreneurs make mistakes to take the market back from them.

3.2. Study of the follow-up strategy for defensive innovation

A large number of empirical studies have shown that most industrial enterprises with more R&D investment tend to implement a follow-up strategy for defensive innovation. The focus of innovation is corrective innovation of existing products, processes, and technical services. A defensive innovation strategy is a better strategic choice for enterprises in declining industries like the food industry.

Founded in 1987, Hangzhou Wahaha Group has developed from a school-run factory into the largest food and beverage company in China within only a dozen years. This feat is impossible without its innovation strategy. At present, Wahaha is equipped with an annual production capacity of 5 million tons of beverages, with supporting production capacity for cans, bottles and caps. It mainly produces six categories of products, including milk-containing beverages, bottled water, carbonated beverages, hot-filled beverages, canned food, medicine and healthcare products, with more than 30 varieties. The sales of bottled water, milk-containing beverages and canned eight-treasure porridge have topped the list in the country for many years. The continuous launch of new products over the years reflects Wahaha's consistent strategy of defending and waiting for opportunities. It is the magic weapon for Wahaha to win the market.

Strictly speaking, most of the products launched by Wahaha are follow-up imitations, which save a lot of costs in the early stage, reduces market risks and improves the success rate of new product launches. There are three ingredients to its success. First, make innovation through imitation; do not to be the first innovator, but follow up the first innovators closely and surpass them. The first product developed by Wahaha was nutrient solution for children. At that time, there were more than 30 companies producing nutrient solution in China, but none of them were targeted at children. Wahaha focused on this market segment and developed the selling point of "appetizer", with the emotional appeal evoked by a bottle of Wahaha as a perfect appetizer to encourage interactions between adults and children. AD calcium milk was first introduced by Robust, and Wahaha followed it up and added the concept of "absorption". Wahaha's tea products are the result of following up Master Kang and Uni-President. While the first movers only promoted the commonality of these kinds of products, Wahaha skipped it and emphasized its individuality as "water from paradise and tea from Dragon Well". For the "Feichang" series, Feichang Coke followed up Coca-Cola and Pepsi for the male market, Feichang Lemon imitated Sprite for the female market, and Feichang Orange mimicked Fanta for the children market. In the early stage of marketing, the series avoided the core market of Coca-Cola — the urban market — and adopted the "rural route", which is a judo strategy. Second, grasp the timing for launch; release products when a mass market is formed. Third, pay attention to speed. Coca-Cola admits that it is slower than Wahaha in market promotion, which is attributed to Wahaha's network advantages as well as unified, centralized organizational structure and decision-making mechanism. These allow it to quickly form advantages of scale while rapidly launching products and then turn them into cost advantages and competitive edges.

4. Institutional Analysis of the Dilemma of Indigenous Innovation

4.1. *Analysis of the dilemma of indigenous innovation*

There are three ways for modern economic growth: accumulation of factors by adding one or more of the three major input factors, including capital, labor and natural resources; transfer of economic structure by transforming industrial structure from low level to high level; technological change. Among factor endowments of a country, natural resources are given and are difficult to increase on a large scale in production. The growth of labor force is limited by birth rate. There is little difference in the changes in labor force across countries, with an annual growth rate of 1% to 3%. The only factor that has a greater impact on economic growth is the change in accumulation of capital. The main difference in the increase in factor inputs among countries lies in the capital accumulation rate. Economic growth can also be achieved through an economic structural upgrading. Reallocation of input factors from inefficient sectors to the efficient ones increases the output from the same number of input factors. Among the three major sources of economic growth, technological change is most critical. Technological innovation prevents the marginal efficiency of capital from declining, thus maintaining long-term economic growth. The extensive expansion of reproduction simply relying on the increase in factor input will sooner or later stagnate. Market competition will deprive this economic growth model of vitality long before resources are exhausted. The upgrading of industrial structure will promote economic growth, but the basis of such upgrading is technological progress. Technological progress allows production of a larger number of more competitive products with fixed factor input, thus promoting the upgrading of industrial structure to achieve economic growth. Technological change is the driving force behind economic development. There are two ways to achieve technological progress: investing in R&D, and learning from and imitating other countries or purchasing advanced technologies from other countries to achieve domestic technological progress. The feature of technological R&D is that the success rate is very low. Generally speaking, 95% of research investment yields no results. Among the technologies with achievements, only a small proportion has commercial value. Hence, the investment in developing new technologies is huge, with a high probability of failure. In contrast, the cost of imitating and purchasing technologies is much lower. Developed countries are at the forefront of technology, so they must achieve technological progress by engaging in R&D. Thus, these countries pay high costs for technological progress, with slow overall progress. Developing countries like China, whose technologies are a far cry from those of the developed countries, enjoy second-mover advantage and can achieve technological progress by means of imitation and purchase. Many studies prove that, even if purchase of patent is adopted, the technological cost is only about one-third of the original cost of development. Moreover, the technologies purchased must have been proved to be successful and commercially valuable, so as to avoid huge losses caused by the failure of R&D. The Japanese economy has maintained rapid growth for nearly 40 years from the 1950s to the 1980s. The Asian Tigers have also maintained rapid growth for nearly 40 years since the 1960s. Their rapid growth is considered to be a miracle. They rely on the technological gap between them and developed countries, and achieve rapid technological progress and rapid economic transformation and growth by introducing technologies.

The problem is that there are certain limitations to the introduction of technologies for economic growth. First, the so-called advanced technologies introduced from abroad are often those that have been eliminated. Relatively speaking, the purchase cost is not low. Second, the long-term economic growth through the introduction of technologies may make its own growth excessively dependent on external development. Thus, the growth momentum of economy has been greatly weakened in the long run. Therefore, indigenous innovation has been essential for the improvement of international competitiveness and long-term economic growth. There are three views on the relationship between technological research activities and technological innovation in the international economic community. According to the first view, the effective way for market economy to allocate research resources is to use them in technological innovation to save the increasingly scarce factors of production. The second view is that the effective way for market economy to allocate research resources is to apply them to technological innovations whose products have a larger market demand. The technological innovation rate of a kind of commodity is a response to the market demand for the commodity. The third view is that a decentralized public research system can effectively allocate research resources in accordance with the two hypotheses of technological innovation above. There have been many studies in the international economic community testing these three hypotheses in the market economy.

Regardless of the above viewpoints, institution plays an important role in innovative activities. Good institutional arrangements will reduce the cost of technology's response to demand. From the first point of view, the benefits from production factors that belong to scarce resources tend to be higher. However, if it is regulated that these kinds of resources can only be produced, studied and developed by a small number of enterprises, the degree of market monopoly will be higher, thus not conducive to technological innovation. The second view is also facing a similar situation. In the third view, the decentralized public research system is only the combination of the above two situations of market demand orientation for technological innovation. Therefore, how to reduce the cost of indigenous innovation through institutional arrangement is an important factor behind the improvement of the level of innovation activities.

4.2. *Dilemma of indigenous innovation and institutional innovation*

Indigenous innovation is an important means for enterprises to seek survival and development as well as enhance their competitiveness. The premise of indigenous innovation is institutional innovation. In his 1990 book *The Competitive Advantage of Nations*, the well-known economist Porter regards the capabilities of innovation and development as the most important factors for a nation to achieve economic success. He further summarizes the perfect economic system as the premise for indigenous innovation. Historical evidence has proved that the economic ups and downs of countries in the modern world are invariably linked to technological innovation and the premise of advantage in such innovation is institutional advantage. The 17th century, British religious reform, the free market system and the secularization of intellectual life gave birth to the first industrial revolution dominated by navigation, energy and communication. It secured Britain the dominant position in the world economy for a century and a half. Religious fanaticism in countries like Spain and Italy stifled innovation. By the first half of the 20th century, the United States had developed a novel professional management system and financial market system that allowed it to quickly surpass Britain in technological innovation. Since the 1990s, technological innovation in different countries has essentially been the comparison of transformation of technological systems and industrial policies. For example, Japan's industrial policies focus on basic sectors, such as new materials, software, environment, and biology. The United States has implemented non-exclusive measures for information and communication technology, such as full openness and cancellation of license management. The result is that the information industry continues to prosper in the United States. The reason for that, first of all, is the enterprise itself. However, we cannot attribute all the reasons to the enterprise. The institutional dilemma that it faces is also an important factor affecting the internal impetus for innovation of the enterprise. The institutional dilemma faced by Chinese enterprises in indigenous innovation is mainly shown in the following three aspects:

First, the subject of indigenous innovation is not clear. The primary element of institutional innovation is the *subject*. The subject of indigenous innovation should and can only be enterprises. In the modern era, new discoveries in science often precede original innovation of technology and the development of applications for industries and processes. However, any kind of technological innovation cannot be transformed into a large-scale industry if it is not operated by enterprises. Nor will it truly become the primary productive force or be converted from knowledge and technology to material wealth. It is even impossible to complete the value circle from inputs through new knowledge and technology to greater material wealth. So, from this perspective, whether enterprises consciously become the subject of innovation is the key to completing the entire industrial innovation process. However, for many years, the Chinese governmental resources have mainly supported colleges, universities and research institutions. The government has offered little financial support for indigenous innovation of enterprises. In some national plans that cover the development of industrial technologies in China, the main subjects of project implementation are still colleges, universities and research institutions. Although participation by enterprises is encouraged, they are often overlooked. Even if some of the planned projects have yielded the expected key technologies, they still fail to enable enterprises, the subject of innovation, to acquire indigenous innovation capacity through the implementation of national plan projects. This situation has left the status of enterprises as the subject of innovation uncertain for a long time. Without the status as the subject or a sound system, indigenous innovation has become a castle in the air.

Second, there are no sufficient incentives for indigenous innovation. The second element of institution is *incentives*. Indigenous innovation depends on an effective market system. The fundamental responsibility of market is to stimulate innovation through competition. Protection for intellectual property rights is a kind of rule in market operation. The nature of such protection is actually the paid spread of technologies on the premise of protecting the interests and enthusiasm of innovators. Without protection, intellectual property rights cannot acquire value in the market. Then no one is willing to innovate, or engage in research, for development and investment. Without protection for intellectual property rights, more people will rely on imitation due to the lack of motivation to engage in indigenous innovation. On the one hand, using the intellectual property system to encourage and promote technological innovation means encouraging innovation by enterprises. Hence, it is necessary to help enterprises gain benefits and compensation. On the other hand, it is also necessary to motivate the excellent professionals for technological innovation with rational remuneration policies. However, most of the innovative professionals in China are not found in enterprises. It is difficult for enterprises to attract innovative talent without innovation incentive mechanisms. Therefore, independent innovation of Chinese enterprises must also be equipped with incentive policies for talent.

Third, there is no policy support for indigenous innovation. The third element of institutional innovation is *policy*. Indigenous innovation should stimulate compatible policies for technological innovation and economic assistance. Enterprises are the subject of innovation and the government has the function of policy guidance. Only by understanding the substantive meaning of policies from the perspective of institutional construction and clarifying the role of the government and the market can policies be more effective. For a long time, Chinese enterprises have difficulty in financing during indigenous innovation, their investment is insufficient, their status as subject is difficult to establish, and they lack technological talent. These problems are closely related to the absence of effective support and efficient cooperation from our economic and technological policies. The central and local governments of China have issued a large number of policies and measures related to technological innovation since reform and opening-up (according to incomplete statistics, the central government alone has issued more than 2,500 policies related to technological innovation). However, these policies are formulated by different departments in different periods for different problems and the degree of evolution of policy issues is different. As a result, there is repetition, gaps, and contradictions of policies for promoting indigenous innovation. They feature inconsistent signals, unscientific design, more call yet less operability, ineffective cost budgeting and control. These have resulted in the lack of systematic policy support for technological innovation in China and the absence of a complete policy system.

Chapter 6

Knowledge Spillover Effect of International Flow of R&D Capital on Indigenous Innovation in China

The enhancement of indigenous innovation capacity depends on the domestic R&D capacity and absorption of knowledge spillover from foreign R&D capital. China has invested far less in R&D than developed countries in the West over the years. Sweden, Finland and Japan had the highest R&D-to-GDP ratio in 2007, reaching 3.73%, 3.45% and 3.39%, respectively. The expenditure on R&D of the United States accounted for 2.68% of GDP, while that of China was 1.44%. In 2008, the United States spent 343.7 billion dollars on R&D in absolute terms, which far outnumbered the amount spent by China (OECD, 2009). When the domestic overall capabilities of R&D activities are limited, absorbing the spillover effect of overseas R&D is particularly essential.

China has to improve its indigenous innovation capacity as characterized by TFP for long-term economic growth. This requires China to gradually increase its capacity to absorb foreign R&D spillovers along with R&D capacity. In fact, foreign R&D capital has directly or indirectly influenced indigenous innovation in China via various channels. Then, do their knowledge spillovers promote China's indigenous innovation? How do the channels of such spillovers, except FDI and import trade, influence indigenous innovation? These urgent questions have to be addressed before China can properly absorb the knowledge spillover from foreign R&D, enhance indigenous innovation capacity and deliver long-term economic growth.

1. Literature Review on Knowledge Spillover of Foreign R&D Capital

1.1. *Spillover effect of import trade*

Coe and Helpman (CH, 1995) were the first to examine the spillover of international trade. Later, researchers improved the CH model in two respects. First, concerning measurement techniques, researchers used dynamic ordinary least squares (DOLS) and found that the coefficient of elasticity of foreign R&D capital stock was no longer robust and not significant statistically. Müller and Nettekoven (1999) tested the CH fixed effects model and proposed a random effect model instead. The second was the improvement in the perspective of selecting weighting. Lichtenberg and Van Pottelsberghe (1996) believed that there is a potential aggregation bias in the CH weighting scheme for foreign R&D capital stock, because a merger between countries may bring up the stock that generates a spillover effect upon aggregation, even if the R&D capital stock and trade volume of the trading partners remain the same. They also used GDP in place of the weight of trade volume in the CH model to remove the potential aggregation bias and drew similar conclusions. The method has now become a mainstream analytical framework since it is not sensitive to data aggregation (Gao and Wang, 2008).

Regarding the spillover effect of import trade, there is also domestic literature comparing the differences in elasticity of productivity of R&D capital between home and abroad by controlling domestic R&D and international R&D capital stock that generates a spillover effect following the method by Lichtenberg and Van Pottelsberghe (1996). The conclusion shows that the elasticity of productivity of the related domestic R&D capital stock is negative. For example, Li (2006) used FFG's six methods to calculate foreign R&D capital stock and analyzed the existence of international R&D capital through the international trade spillover. The research offers evidence that the coefficient of the industry's technological progress affected by R&D capital of the industry is significantly negative.

1.2. *Research on the existence of interregional knowledge spillover*

Regarding knowledge spillover between regions, Romer (1986) argued that knowledge is non-competitive and partially exclusive, which leads to the emergence of knowledge spillover effects. Caniels and Keilbach (2000) established a multiregional model of economic growth driven by knowledge spillover. Grossman and Helpman (1995) argued that knowledge spillover in one region can drive economic development in other regions. Griliches (1986) reviewed studies on R&D spillover effects and pointed out that R&D spillover effects are universal and play an important role in economic growth. Anselin *et al.* (1997) used spatial econometric analysis to consider the geospatial influence of economic factors. They introduced spatial lag to extend the production function and carry out an empirical study on R&D spillover based on data on R&D and high-tech innovation of American universities in 1982. They found that R&D spillover transcends geographic boundaries. Funke and Niebuhr (2000) tested R&D spillover effects with the data on the former West German region for the period between 1976 and 1996 and reached similar conclusions. Although the methods and data used by scholars are different, they all find that R&D spillover effects are indeed universal and contribute significantly to economic growth in neighboring regions.

Few of the Chinese scholars have investigated the relationship between R&D spillovers and economic growth. Wang (2003) tested the influence of R&D spillovers on economic growth with data on China and the United States for the period between 1990 and 1999 based on the two-country Mundell–Fleming model in the open economy. It was found that R&D spillover of the United States has a significant effect on China's GDP growth. Su (2006) conducted spatial econometric analysis on the spatial scope and degree of R&D knowledge spillovers of China. It was found that R&D knowledge production has spatial dependence. The production of R&D knowledge in one region not only increases its stock of knowledge but also spills over to the adjacent regions, causing the increase in their stock of knowledge. Wu (2004) adopted a spatial error model that considers spatial autocorrelation and conducts a spatial econometric study on the convergence of provincial economic growth in China. It was found that the spatial connection of provincial economy in China has been continuously strengthened and that both geographical factors and spatial effect have had an important impact on economic growth. Zhang *et al.* (2007) used data on 31 provinces in China in 2004 and conducted a spatial econometric analysis of the number of provincial patent grants and the association mechanism between three types of patent grants and regional economic growth. They found that patent innovation and its three forms make a significant contribution to regional economic growth.

Patent innovation in one region has spillover effects, which contribute to the economic growth in neighboring regions.

According to existing studies, most of the research on the channels of overseas R&D capital and knowledge spillovers is limited to FDI and import trade, which is not systematic or comprehensive enough. More importantly, existing studies do not take the measurement of overseas R&D capital as the research premise, but directly carry out tests with FDI or the amount of import trade of the current year as an explanatory factor. This obviously leads to certain deviation. According to Mohnen (2001), there are six channels to realize international technology spillovers, namely, FDI, international trade in goods, immigration, publication of inventions and patent transfer, international cooperation and mergers and acquisitions, and purchase of foreign technologies. Existing theoretical and empirical analyses of the channels of R&D spillovers are generally conducted from the perspective of a certain channel and mainly focus on FDI and import trade. They rarely consider the technology spillover effects of other channels, nor do they integrate multiple channels into one model. In this book, we intend to expand and optimize the explanatory variables and their weight selection under the knowledge spillover model established by Coe and Helpman. There are direct and indirect channels of the knowledge and capital spillovers. We consider the impact of overseas R&D capital and knowledge spillovers on indigenous innovation under various channels and use overseas R&D capital as the premise of knowledge spillover. We select 24 representative countries with high investment in R&D, measure the stock of overseas R&D capital and calculate the knowledge spillover effect of each channel based on their weight. On such a basis, we measure the contribution of five spillover channels, including FDI and import trade, to indigenous innovation in China.

2. Extension of Knowledge-Driven Endogenous Growth Model

2.1. *Explanation of basic models*

Based on a knowledge-driven endogenous growth model, we draw on the regression methods of CH (1995) and Lichtenberg and Van Pottelsberghe (1996) to analyze knowledge spillover. We assume that a country's TFP is related to the R&D capital of the country as well as others. TFP is defined as

$$\text{TFP} = AS^\lambda, \tag{1}$$

where A is a constant, representing the exogenous economic factors, and S is the knowledge capital, i.e., the accumulation of R&D investment of an economy. For an open economic system, S depends not only on S^D, the accumulation of domestic R&D investment, but also on S^F, the accumulation of R&D investment in overseas countries. Thus, under open economic conditions, R&D capital can be defined as

$$S = (S^D)^\alpha (S^F)^\beta. \tag{2}$$

Among them, the transnational spillover effect of the accumulation of R&D investment of overseas countries can be divided into direct spillover and indirect spillover based on how the TFP growth of host countries is affected. The level of direct transnational spillover of overseas R&D capital is S^{FD} and that of indirect transnational spillover is S^{FI}. Thus, we define overseas knowledge capital as

$$S^F = (S^{FD})^\gamma (S^{FI})^\chi. \tag{3}$$

2.2. Expansion of basic models

According to the above basic models, we further distinguish direct and indirect spillover effects of overseas R&D capital. Direct effect is divided into direct introduction of technologies from abroad and direct R&D investment from MNCs in China. Obviously, both direct introduction of technologies and direct R&D investment will be applied to the technological innovation activities of enterprises in China. Therefore, these two types of activities also have a direct impact on TFP growth. On the basis of this, we further define the cross-border direct spillover of overseas R&D capital as

$$S^{FD} = (S^{TIC})^\delta (S^{FRD})^\varphi, \tag{4}$$

where S^{TIC} is the capital stock of technology contracts introduced from abroad and S^{FRD} is the stock of overseas direct R&D investment in China.

As regards the indirect spillover of overseas R&D capital, we believe that there are three channels. First, Chinese enterprises carry out imitative innovation by importing finished products and products in progress from abroad, thereby improving their own technological level. Considering that the spillover effects of imported capital goods and consumer goods may be different, this book further divides the import channels into capital goods import channels and consumer goods import channels. Second, Chinese enterprises enhance their technological level through the competition effect and the talent flow effect by virtue of the investment behaviors of FDI enterprises in China. Third, Chinese enterprises make FDI. During overseas investment, they hire local researchers and compete with local enterprises, which will enhance their technological level in the home country (China). On the basis of this, we further define the cross-border indirect spillover of overseas R&D capital as

$$S^{\mathrm{FI}} = (S^{\mathrm{IMP\text{-}C}})^{\phi} (S^{\mathrm{IMP\text{-}K}})^{\omega} (S^{\mathrm{FDI}})^{\eta} (S^{\mathrm{OFDI}})^{\mu}, \qquad (5)$$

where $S^{\mathrm{IMP\text{-}C}}$ is the foreign R&D capital spillovers absorbed when consumer goods are imported from other countries, $S^{\mathrm{IMP\text{-}K}}$ is spillovers absorbed when capital goods are imported from other countries, S^{FDI} is spillovers absorbed when other countries make a direct investment in China and S^{OFDI} refers to the R&D capital spillovers absorbed when Chinese enterprises make a direct investment in other countries.

If we substitute Eqs. (5), (3) and (4) into Eq. (2) and combine the result of Eq. (1), then

$$\begin{aligned} \mathrm{TFP} &= f(S^{\mathrm{D}}, S^{\mathrm{F}}) = f(S^{\mathrm{D}}, S^{\mathrm{FD}}, S^{\mathrm{FI}}) \\ &= f(S^{\mathrm{D}}, S^{\mathrm{TIC}}, S^{\mathrm{FRD}}, S^{\mathrm{IMP\text{-}C}}, S^{\mathrm{IMP\text{-}K}}, S^{\mathrm{FDI}}, S^{\mathrm{OFDI}}). \end{aligned} \qquad (6)$$

2.3. Measurement of indirect spillover channels of overseas R&D capital

At present, measurement of spillover channels in the academic community is mostly made by using the current amount of FDI or import trade as an indicator for the test, and there are large deviations. The amount of investment and trade can only characterize the development of the two channels; it is impossible to identify the flow of overseas R&D capital into China through the two channels. Overseas R&D capital comes from different countries and regions. Each region differs in output of technology contracts, direct R&D investment to China, export to China, direct investment in China and reception of direct investment from China. Besides, in terms of import channels, the spillover effects of imported capital goods and consumer goods are also distinctly different. Therefore, we will recalculate the flow of overseas R&D capital into China through these channels. To simplify the calculation based on the data available, we assume that overseas R&D capital spills over equivalently through import, FDI and OFDI. At first, we calculate the spillover effects of each of the countries or regions through the above three indirect channels. Then, we work out the sum to obtain the overall spillover effect of each channel in China. Take the import channels of capital goods as an example. Initially, the spillover effect through the import channel of capital goods in country j is calculated. i.e.,

$$S_j^{\text{IMP-K}} = \frac{\text{IMP-K}_j}{\text{IMP-K}_{j\text{-total}}} S_j^{\text{D}},$$

where S_j^{D} is the domestic R&D capital stock of j, IMP-K$_j$ is the total value of capital goods that j exports to China (the total value of capital goods that China imports from j, thus called the import channel of capital goods) and IMP-K$_{j\text{-total}}$ is the total export value of capital goods of j in the same year. Then, add up spillovers through the import channel of capital goods of different countries and obtain

$$S^{\text{IMP-K}} = \sum_{j=1}^{n} \frac{\text{IMP-K}_j}{\text{IMP-K}_{j\text{-total}}} S_j^{\text{D}}.$$

Similarly, the spillover effects through import channel of consumer goods as well as channels of attracting foreign investment and making foreign investment can be calculated. i.e.,

$$S^{\text{IMP-C}} = \sum_{j=1}^{n} \frac{\text{IMP-C}_j}{\text{IMP-C}_{j\text{-total}}} S_j^{\text{D}}, \quad S^{\text{FDI}} = \sum_{j=1}^{n} \frac{\text{FDI}_j}{\text{FDI}_{j\text{-total}}} S_j^{\text{D}}$$

$$\text{and} \quad S^{\text{OFDI}} = \sum_{j=1}^{n} \frac{\text{OFDI}_j}{\text{OFDI}_{j\text{-total}}} S_j^{\text{D}}.$$

In addition, we will carry out empirical analysis based on panel data. Using the above method of calculation, we make an evaluation based on the proportion of each province. The spillover effect of overseas R&D capital obtained by i through import of capital goods is

$$S_i^{\text{IMP-K}} = \frac{\text{IMP-K}_i}{\text{IMP-K}} \sum_{j=1}^{n} \frac{\text{IMP-K}_j}{\text{IMP-K}_{j\text{-total}}} S_j^{D},$$

where IMP-K_i is the total import value of capital goods of i and IMP-K is the national total import value of capital goods over the year. The knowledge spillover effect of overseas R&D capital obtained by i through channels of consumer goods is

$$S_i^{\text{IMP-C}} = \frac{\text{IMP-C}_i}{\text{IMP-C}} \sum_{j=1}^{n} \frac{\text{IMP-C}_j}{\text{IMP-C}_{j\text{-total}}} S_j^{D},$$

where IMP-C_i is the import value of consumer goods of i over the year and IMP-C is the national import value of consumer goods over the year. The knowledge spillover effect of overseas R&D capital acquired by i through the FDI channel is

$$S_i^{\text{FDI}} = \frac{\text{FDI}_i}{\text{FDI}} \sum_{j=1}^{n} \frac{\text{FDI}_j}{\text{FDI}_{j\text{-total}}} S_j^{D},$$

where FDI_i is the stock of FDI utilized by i over the year and FDI is the stock of FDI utilized by China. The spillover effect obtained by i through channel of making foreign direct investment is

$$S_i^{\text{FDI}} = \frac{\text{FDI}_i}{\text{FDI}} \sum_{j=1}^{n} \frac{\text{FDI}_j}{\text{FDI}_{j\text{-total}}} S_j^{D} \frac{\text{ODFI}_j}{\text{OFDI}_{j\text{-total}}} S_j^{D},$$

where OFDI_i is the stock of FDI made by i over the year and OFDI is the stock of FDI made by China over the year.

3. Empirical Analysis of Knowledge Spillovers of Overseas R&D Capital

3.1. *Model and data*

According to the theoretical derivation and the calculation of related channels in the previous section, we draw on the CH model and set the regression equation for the knowledge spillover effect of overseas R&D capital on China's indigenous innovation as

$$\ln(\text{TFP}_{it}) = a_i + \beta_1 \ln(S_{it}^D) + \beta_2 \ln(S_{it}^{\text{TIC}}) + \beta_3 \ln(S_{it}^{\text{FRD}}) + \beta_4 \ln(S_{it}^{\text{IMP-C}})$$
$$+ \beta_5 \ln(S_{it}^{\text{IMP-K}}) + \beta_6 \ln(S_{it}^{\text{FDI}}) + \beta_7 \ln(S_{it}^{\text{OFDI}}) + \varepsilon_{it}. \tag{7}$$

By observing the regional distribution of overseas R&D capital and the major import and export trading partners, FDI source countries and destination countries of China's FDI, we select 24 countries as source countries of overseas R&D capital, which mainly include OECD countries, such as USA, Japan, UK, France, Germany, Canada, Sweden, Turkey, Austria, Belgium, Czech Republic, Denmark, Finland, Greece, Ireland, the Netherlands, Norway, Portugal, Spain, South Korea, Hungary and Poland, as well as non-OECD countries, like Singapore and the Russian Federation. These countries are representative as they are major import and export trading partners of China, make direct investments and R&D investments in China, or are recipients of China's FDI. They take up an absolute proportion of the amount for each item.

In Eq. (7), S_{it}^D is the knowledge capital stock of i in the tth year. There are two kinds of measures for R&D activities, namely, amount of R&D investment commonly used in the academic community (Keller, 1998; Xie and Zhou, 2009) and the R&D input density adopted by Mansfield (1981) to circumvent the impact of differences in economies of scale on R&D activities. Based on the data available, we choose R&D input of each region as a measure since we are mainly calculating the R&D capital stock in this book. The perpetual inventory method is used to calculate R&D capital stock. With 1992 as the base year, we draw on the method of Griliches (1992) and divide the R&D investment flow in 1992 by the depreciation rate and the average growth rate of the 5 years following the base year as the R&D stock of 1992, since there are no detailed data. That is, $S_{i1992}^D = R_{i1992}/(\delta + \zeta)$, where R_{i1992}^D is the R&D expenditure of i in 1992, δ is the depreciation rate, generally set at 9.6%, and ζ is the average growth rate of the 5 years following the base year. On the basis of this, the perpetual inventory method is used. That is, R&D stock in t = R&D stock in $t - 1 \times (1-9.6\%)$ + R&D flow in t (unless otherwise stated, the stock of the base year and other years is calculated in the same way as S_{it}^D). The basic data for calculation of S_{it}^D are the R&D expenditure of each region derived from the *China Statistical Yearbooks on Science and Technology* from 2004 to 2008.

S_{it}^{TIC} is the capital stock of technology contracts introduced from abroad in the tth year of i. With 1992 as the base year, it is calculated based on the amount of technology contracts introduced to various provinces from abroad. Then, the perpetual inventory method is used to calculate the stock of other years. Since overseas capital tends to have a lower depreciation rate than domestic capital, the depreciation rate is set at 5%. The basic data for calculation of S_{it}^{TIC} are the amount of contracts for technology introduction of each region derived from the regional statistical yearbooks from 2004 to 2008.

S_{it}^{FRD} is the stock of direct R&D investment in i by other countries in the tth year. The stock is calculated based on the R&D capital flows absorbed by the provinces. The flows consist of foreign R&D funds attracted by R&D institutions and those absorbed by large and medium-sized enterprises (three types of foreign-funded enterprises). Considering the availability of data, we use 1994 as the base year and measure the stock of direct R&D investment in other years using the perpetual inventory method, with a depreciation rate of 5%. The basic data for calculation of S_{it}^{FRD} are the foreign R&D funds absorbed by R&D institutions in the regions and those by large and medium-sized enterprises (three types of foreign-funded enterprises) in the regions, derived from *China Statistical Yearbooks on Science and Technology* from 2004 to 2008.

$S_{it}^{\text{IMP-C}}$ and $S_{it}^{\text{IMP-K}}$ are the R&D capital stocks of other countries absorbed by i through import of consumer goods and capital goods, respectively. The method mentioned above also applies here. The capital goods and consumer goods are classified based on by the standards for classification of goods in international trade Broad Economic Categories (BEC). Under the BEC issued by the United Nations Statistics Division, capital goods include machinery, other capital goods (excluding transportation equipment) and industrial transportation equipment; consumer goods include primary food and beverages as well as processed food and beverages for household consumption, non-industrial transportation vehicles and other consumer goods not specified. During data processing, data on capital goods and consumer goods that China imports from j were taken from the conversion and calculation of import of capital and consumer goods under BEC using the SITC standards established by UNCTAD. Import of i in China is based on conversion and calculation of import of capital and consumer goods under BEC according to the classification of custom goods. We refer to Statistical Papers Series M No. 53, Rev.4 of the United Nations (M/53/Rev.4) for the method of conversion. Data on capital goods and consumer goods that China imports from j as well as the total export of capital goods of j are derived from UNCTADstat. Data on total import of consumer goods and capital goods import of China and that of i are derived from the China Economic Database.

S_{it}^{FRD} refers to the stock of overseas R&D capital obtained by i through FDI. The calculation method mentioned above also applies here, based on the FDI flow of the provinces, with 1992 as the base year and 5% as the depreciation rate. Data on j's investment in China and its total FDI are derived from UNCTADstat. Data on overall FDI of China and that of i are derived from the *China Statistical Yearbook*s from 2004 to 2008.

S_{it}^{OFDI} refers to the R&D capital stock of other countries that i absorbs through FDI. The calculation method mentioned above also applies here, based on the FDI flow of the provinces, with 1992 as the base year and 9.6% as the depreciation rate. Data on FDI of i and that of China and the direct investment of China in j are derived from the China Foreign Investment Bulletin (2004–2008). Data on the total FDI inflow of j are derived from UNCTADstat.

S_{jt}^{D} is the domestic stock of R&D capital of j in the tth year. It is calculated based on the R&D expenditure flow of the countries, with 1992 as the base year and 5% as the depreciation rate. Data on the domestic R&D expenditures of the 24 countries are derived from the *Main Science and Technology Indicators in the OECD Countries* (2004–2008). The unit "US dollar" in the above data is converted into RMB based on the average Chinese yuan exchange rate against the US dollar in the current year.

In Eq. (7), TFP_{it} is the TFP of i in the tth year, calculated according to the Malmquist index of DEA. The input factors are the capital stock K and the number of employees L in each region and the output indicator is the GDP of each region. The capital stock is measured by the gross fixed capital formation in the current year of each region and the number of employees by the total number of employees in the three sectors of each region. GDP is the gross domestic product of each region. Among these factors, K and GDP are deflated based on the corresponding price index and converted into the constant price of 1990. All the data are derived from the *China Statistical Yearbook*s from 2003 to 2008.

Table 1. Performance of TFP of China from 2003 to 2007 (Malmquist Index)

Year	EFch	TFch	PEch	SEch	TFPch
2003	0.976	1.036	1.015	0.962	1.011
2004	0.976	1.104	0.989	0.978	1.068
2005	1.031	1.052	1.035	0.995	1.085
2006	1.005	1.084	0.989	1.016	1.089
2007	1.091	1.031	0.996	1.095	1.125
Mean value	1.014	1.061	1.005	1.009	1.075

3.2. *Measurement and decomposition of China's TFP*

We use DEAP 2.1 software for DEA analysis and measure the TFP, technological efficiency and technological progress of each region from 2003 to 2007. The results are shown in Table 1.

According to Table 1, it is obvious that TFP is increasing steadily from 2003 to 2007, with an average growth rate of 2.7%. Technological efficiency, except a decline in 2006, is on the rise, with an average growth rate of 4.2%. Yet there are some fluctuations in technological progress (TEch). Considering that TFP can measure the contribution of input factors, other than capital and labor, to economic growth, we use TFP as an explanatory variable to study the contribution of overseas R&D capital to China's indigenous innovation. In addition, since observation of the changes in technological efficiency and technological progress allows an in-depth and detailed view of the changes in TFP, we also consider the changes in technological efficiency (EFch) and technological progress change (TEch) based on Eq. (7) as explained variables for comparative study.

3.3. *Analysis of knowledge spillover effects of overseas R&D capital*

We examine samples of 30 provinces in China (excluding the Tibet Autonomous Region, Hong Kong SAR, Macao SAR and Taiwan Province due to data availability) and analyze the impact of different spillover channels on China's indigenous innovation. The descriptive statistics for each variable are presented in Table 2.

Table 2.   Descriptive Statistics of Variables

Variable	Sample size	Mean value	Standard deviation	Minimal value	Maximal value
ln TFP	150	0.008	0.034	−0.098	0.128
ln EFch	150	0.005	0.049	−0.039	0.097
ln TEch	150	0.002	0.087	−0.108	0.136
ln SD	150	6.127	0.595	4.747	7.325
ln STIC	150	4.964	0.813	2.000	6.565
ln SFRD	150	6.373	0.592	4.924	7.409
ln SIMP-C	150	3.788	0.607	1.781	5.006
ln SIMP-K	150	3.329	0.864	1.129	5.516
ln SFDI	150	11.168	0.606	9.158	12.366
ln SOFDI	150	10.709	0.864	8.506	12.893

Table 3.   Correlation Coefficient Matrix of Variables

	ln SD	ln STIC	ln SFRD	ln SIMP-C	ln SIMP-K	ln SFDI	ln SPFDI
ln SD	1	—	—	—	—	—	—
ln STIC	0.1386	1	—	—	—	—	—
ln SFRD	0.0530	0.1557	1	—	—	—	—
ln SIMP-C	0.1020	0.0724	0.0673	1	—	—	—
ln SIMP-K	0.0758	0.1493	0.1961	0.2202	1	—	—
ln SFDI	0.1033	0.0596	0.1392	0.1291	0.4197	1	—
ln SOFDI	0.2697	0.1474	0.1973	0.2196	0.996	0.2203	1

Considering the important role of foreign-invested enterprises in the import of China and the possible correlation between other variables, we measure the correlation coefficients between variables, as shown in Table 3.

As the matrix reveals, the correlation coefficients between most variables are below 0.2. But it is relatively obvious that the correlation coefficient between FDI and SIMP-K reaches 0.4197. This is mainly because a large amount of capital in China is imported by foreign-invested enterprises. This will be further discussed when we carry out the empirical test.

We perform regression analysis for the period from 2003 to 2007 with Stata10.0. To eliminate the influence of cross-sectional heteroscedasticity,

Table 4.   Regression Results of R&D Spillover Model

Variable	Model 1 ln TFP as explained variable	Model 2 ln EFch as explained variable	Model 3 ln TEch as explained variable
ln SD_{t-1}	0.125*	0.004*	0.050***
ln STIC	0.019	0.103	−0.036
ln SFRD	0.061	0.026**	0.032
ln SIMP-C	−0.023	−0.591	−0.403
ln SIMP-K	0.104	0.061**	0.137**
ln SFDI	0.010	0.014**	0.025*
ln SOFDI	0.084	0.003	0.129
C	−7.86	−10.28	−8.01
Adjusted R^2	0.701	0.524	0.677
F	3.62***	2.88***	2.74***
Hausman	13.91***	10.37***	10.38***

Notes: ***, ** and * refer to passing the test at a significance level of 1%, 5% and 10% respectively.

we use the cross-sectional weighting method. Meanwhile, to avoid possible endogenous problems between R&D investment and TFP, we introduce ln SD_{t-1} into the model. The results of the analysis with the individual fixed effects model based on the Hausman test are shown in Table 4.

As the estimated results of Model 1 in Table 4 indicate, the introduction of technology contracts in the direct channels and the import of consumer goods in the indirect channels do not pass the significance test at the level of 10%. The FDI channel and the direct R&D of MNCs are significant at the level of 10%. The domestic R&D capital passes the significance test at the level of 5% and its regression coefficient is also the largest among the variables. Import of capital goods and FDI in R&D in China also pass a significance test at the level of 5%. FDI shows weaker significance. In addition, existing studies are mostly limited to analysis of the spillover of import trade and FDI channels. In this book, we measure the spillover of overseas R&D capital through import trade of capital goods and FDI channels with a new approach, thus obtaining the regression results of Model 3. As the results suggest, FDI has much weaker spillover effects than import trade of capital goods, which points to the irreplaceable role of import of capital goods in indirect channels.

Based on the above results, it is easy to see that the accumulation of domestic R&D capital and the contribution of five spillover channels to indigenous innovation show the following different characteristics.

- **First, China's domestic R&D capital is the most important determinant of indigenous innovation:**
  As the regression results in Table 4 reveal, domestic R&D capital stock has had a significantly positive effect on TFP as well as changes in technological efficiency and technological progress. There are only some differences in the details of the effect. According to Table 4, local R&D capital mainly influences TFP by affecting the changes in technological progress but has no obvious effect on the changes in technological efficiency. Therefore, local R&D capital can only be taken as a factor affecting China's technological progress rather than improvement in efficiency. Adhering to the strategy of "national rejuvenation through science and education", China has increased investment in science and technology as a major strategic measure to promote indigenous innovation and economic growth. In recent years, it has been emphasizing the role of indigenous innovation. As China increasingly expands its investment in R&D, the direct promoting effect of R&D capital on indigenous innovation is continuously boosted. On the contrary, R&D capital has indirectly enhanced China's absorption of the knowledge spillover of overseas R&D capital by helping to improve technological resources and human capital. In this way, the indigenous innovation capacity is enhanced. To be specific, R&D investment has accelerated the production of new knowledge and technologies that play a decisive role in improving the quality and value added of products as well as content of technology and knowledge. Meanwhile, R&D investment has spawned new products and industries, promoted the upgrading of industrial structure, created new jobs and given rise to a new social division of labor. It has greatly improved the level of production technologies and equipment, so that more output can be produced with fewer resource input. In addition, a large number of scientific and technical professionals have improved their own quality in R&D activities, which has led to an increase in the efficiency of all workers. However, this mechanism of action is restrained by factors such as inappropriate input structure of R&D capital and low efficiency of R&D investment (Li and Zhu, 2006; Xie and Zhou, 2009).

- **Second, import trade of capital goods is a major indirect spillover channel, while that of consumer goods has no significant effect:**
  According to Table 4, knowledge spillover of overseas R&D capital through import of capital goods has a positive effect on TFP,

technological efficiency and technological progress and is significant at the level of 5%. In contrast, import of consumer goods has no significant effect on the three explained variables. Knowledge spillover through import of capital goods mainly plays a role through the mechanisms of competition, correlation and imitation. Faced with imported capital goods that feature better quality and higher technology, local manufacturers must maintain their market power by improving their competitive advantage or innovating on the basis of learning from and imitating the imported products. Importers usually obtain knowledge spillovers while receiving guidance and trainings on advanced technologies and products from foreign exporters. It is worth noting that import of consumer goods has no significant contribution to indigenous innovation. Scholars only paid attention to the spillover effect of the total trade volume on indigenous innovation. However, the existence of consumer goods obviously affects the intensity of the overall spillover effect. Therefore, the prerequisite to a correct understanding of the spillover effect of overseas R&D capital through import is further differentiation between consumer goods and capital goods, which have different effect intensities and channels in terms of indigenous innovation.

- **Third, direct investment in R&D in China is a major direct spillover channel, with significant contribution to technological efficiency:**
  As the coefficient of explanatory variables indicates, knowledge spillover through direct channels of overseas R&D investment has a promoting effect on TFP. The effect is significant at the level of 5%, yet is weaker than import trade of capital goods. Meanwhile, direct R&D investment in China has a significantly positive effect on the changes in technological efficiency but no significant contribution to technological progress. In recent years, an increasing number of MNCs have set up R&D institutions in China to enhance their competitive advantage, or make direct R&D investment in China through cooperation with local universities and research institutes. Their influence on indigenous innovation is mainly generated through four spillover channels, namely, technology infusion mechanism, knowledge flow mechanism, imitation mechanism and personnel mechanism. In general, the knowledge spillover of direct investment in overseas R&D is supposed to have widespread and significant effects on TFP. Yet the

result of our test shows that the effect is not very obvious. We believe this is probably because, for the R&D centers established by MNCs in China, most R&D activities are focused on the localization of foreign advanced technologies centered on the domestic market, rather than R&D of new technologies. In addition, since overseas R&D investment is a new phenomenon in recent years, MNCs tend to cautiously keep key technologies in home countries, or have foreigners control key positions and key technologies in their R&D institutions in China, thus hindering the functioning of personnel mechanism.

- **Fourth, the spillover effect of overseas R&D capital through FDI channels should not be overestimated:**
  As the coefficient of explanatory variables indicates, knowledge spillover of overseas R&D capital through FDI channels has the smallest effect on TFP and limited impact on technological efficiency and technological progress. Thus, the spillover effect of this channel should not be overestimated. However, FDI still has a positive effect on indigenous innovation. For enterprises in the same industry, the entry of MNCs will intensify market competition in host countries and add to the survival pressure of enterprises of host countries. As a result, these enterprises will improve their technologies by continuously increasing their technological input and improving resource allocation, so as to survive and develop in competition, narrow the technological gap between them and MNCs and better absorb the knowledge spillover of overseas R&D capital. On the contrary, local enterprises absorb such spillover by learning from and imitating the advanced products and technologies of MNCs. They then improve their own products and technologies to a certain degree based on market characteristics so as to produce better products and technologies. For upstream and downstream enterprises, under the effect of linkage mechanism, they have forward or backward links with MNCs and obtain knowledge spillover of overseas R&D capital by producing products and services that meet the specific standards of MNCs, receiving trainings from MNCs on products and technologies as well as learning and mastering a large amount of technical and managerial knowledge. In addition, local professionals employed by MNCs have gradually mastered a set of production methods of foreign-funded enterprises, their management techniques, organizational structure and models, and ways to combine production factors like products

and processes in the course of work. This has improved the overall level of human capital of host countries. When these employees quit and work for other companies, this R&D spillover is applied to other enterprises of host countries.

- **Fifth, technology introduction contracts do not play the role of spillover:**
  According to the degree of significance of the explanatory variables, knowledge spillover of overseas R&D capital through technology introduction contracts has no significant effect on TFP or changes in technological efficiency and technological progress and cannot constitute a spillover effect on indigenous innovation. Knowledge spillover of technology introduction contracts is a relatively straightforward way of technology infusion. The effect on domestic indigenous innovation from direct introduction of foreign advanced technologies is mainly generated through three spillover channels, namely, technology infusion mechanism, knowledge flow mechanism and personnel mechanism. However, few works of literature have mentioned the knowledge spillover effect of technology introduction contracts, which has proved to be unobvious in the test conducted in this book. We believe that one of the main reasons is that the transaction amount in the market of technology introduction contracts is small and the acquired knowledge spillover of overseas R&D capital is constrained by the contracts, with limited scope of dissemination and diffusion. Hence, the knowledge spillover fails to have an apparent effect on indigenous innovation and economic growth.

- **Sixth, FDI is not a factor that promotes indigenous innovation:**
  According to the test results, the knowledge spillover of overseas R&D capital through FDI channels has no significant effect on indigenous innovation. One reason is that the motivation behind Chinese enterprises going global is not a quest for technology. Theoretical research shows that the main motivation of reverse FDI of developing countries is to acquire advanced knowledge and technology from developed countries, but this is not very obvious in the case of China. Many enterprises that go international are market oriented (Hou, 2009). In many cases, Chinese enterprises engage in foreign investment for strategic reasons, or for survival and development under the competitive impact of MNCs. They make strategic direct investment

in developed countries to squeeze such countries' market and protect their own market share. Therefore, the knowledge spillovers of China's FDI that is not technology oriented have no obvious positive impact.

4. Summary

Based on a literature review, we explore the knowledge spillover mechanism of overseas R&D capital, derive and extend the R&D spillover model of CH, propose five channels of knowledge spillover of overseas R&D capital and further divide import channels into those for capital goods and those for consumer goods for comparative analysis. In terms of empirical research, we select 24 representative countries with high R&D input, measure the stock of overseas R&D capital, calculate their knowledge spillover effects based on the weight of each channel and carry out empirical research with data on 30 provinces, municipalities and autonomous regions from 2003 to 2007. We find that the knowledge spillovers of domestic R&D capital and overseas R&D capital through import channels of capital goods have a significant and strong positive effect on indigenous innovation, followed by overseas direct investment in R&D and FDI. Import of consumer goods, overseas direct investment and technology introduction contracts have no obvious effect.

With the increase of R&D investment in China, R&D capital has been playing an increasingly important role in directly promoting indigenous innovation. R&D capital has indirectly improved China's ability to absorb the knowledge spillover effect of overseas R&D capital by promoting the improvement of technological resources and human capital. It has exerted the most significant and important impact on indigenous innovation. Among the various spillover channels, the import channel has an obvious effect. One of the major innovations of this book is the distinction between import of capital goods and that of consumer goods. The results reveal that, compared with import of consumer goods, which has no significant effect, import of capital goods shows the greatest effect among all the spillover channels. Direct investment in overseas R&D and FDI also has a positive effect but with limited intensity. Hence, the spillover effect of FDI cannot be overestimated. We believe that direct investment in overseas R&D stems from the R&D centers established by MNCs in China. Most R&D activities are focused on the localization of foreign advanced technologies based on the domestic market, rather than the

R&D of new technologies. This to a certain extent affects the significance level of the spillover effect. The R&D spillover effects of overseas direct investment and technology introduction contracts are not obvious, which has rarely been mentioned in previously published literature. We believe that one of the main reasons is that the amount of China's overseas direct investment and technology introduction contracts is small, with limited scope of spread and radiation. However, with the expansion of overseas investment and the increase in technology introduction, the knowledge spillover of overseas R&D capital will have a more significantly positive effect on indigenous innovation.

Chapter 7

Path Choice for the Enhancement of Indigenous Innovation in China

1. Participation in the Global Alliance of Technology Innovation

A strategic technology alliance is a partnership established by two or more MNCs to achieve a certain strategic goal. Vonortas *et al.* use the statistics of 2,683 companies in the IT industry from 1984 to 1994 to study the impact of strategic technological alliances of MNCs on developing countries. According to the results, the technology spillover effects of strategic technology alliances enable enterprises in developing countries to acquire the advanced technologies of MNCs, thereby improving the technological innovation capacities of local enterprises. The development of economic globalization and the implementation of the R&D internationalization strategy of MNCs have strengthened the connection between countries around the world and changed the living conditions of enterprises in various countries. MNCs[1] assume a leading and core position among the subjects of innovation. Local enterprises in China rely on innovation vitality, professional advantages and understanding of the local market to establish partnership with these MNCs for division of labor and cooperation.

---

[1] In fact, there are also a small number of large enterprises in China that have strong innovation capacity, but the proportion is extremely small. For the convenience of analysis, this will not be pointed out in the following sections.

1.1. M&A or technology alliance with integration of industry chains as the core has become an important means for R&D internationalization of MNCs

Competition between enterprises has evolved into that between alliances as well as that between enterprise clusters. Cooperation based on competition and development in competition are new models of joint competition. In addition to hiring local employees, it is also necessary for the MNCs to further cooperate with local enterprises to "combine the internal with the external", based on the need to increase the understanding of China's domestic market and improve market competitiveness. Many international engineering enterprises cooperate with our local ones in the form of outsourcing cooperation and business subcontracting. Therefore, we will guide multinational R&D institutions to work closely with local enterprises to promote technology transfer and talent spillovers of such institutions, give full play to the "competition based on coopetition" between the two sides, and bring a number of domestic enterprises into the international market. This is the most typical form of alliances of technological innovation. In addition, pressured by cruel market competition, especially the vicious price competition and the mutual plagiarism of technology, domestic enterprises often form such alliances to safeguard and guarantee the interests of all parties.

1.2. With increasing investment in technological innovation and increasingly higher risks, enterprises are forming alliances of technological innovation to reduce the risk of R&D investment

Under the trend of economic globalization, the economic scale and volume are becoming increasingly larger. The models of market competition and the models of corporate competition also experience some changes quietly. The confrontational competition between enterprises in the past has gradually weakened and is transforming into user-centered "coopetition" for mutual benefit and win−win results. With the development of R&D internationalization and strategic technology alliances of enterprises, major MNCs have formed a global network of complex technologies through various kinds of technical contact and, with this network, control global S&T resources. Targeting the global market, they require each other to simultaneously launch homogenous products that meet different consumption habits. It is impossible to achieve these goals by relying on one MNC. These MNCs are also facing disastrous consequences caused by failures in technological innovation. Hence, in a cooperative attitude, they focus on the overall target, work together to tackle problems and take actions, as well as establish a cooperative entity to seek breakthroughs in core areas, thus sharing the huge profits brought about by the success of technological innovation as well as bearing the loss caused by the failures in technological innovation.

1.3. *Combination of technical standards and patented technologies has become a powerful weapon of MNCs in their strategic enclosure movements*

Every time the technological transformation and upgrading have led to the rise of a large number of new enterprises, thus instilling new vitality into social and economic progress. Old enterprises that fail to keep abreast of development or adjust their product mix will be eliminated. Yet, once they seize the opportunity and catch up, they will suddenly emerge. That is to say, each upgrade of technologies and equipment offers an opportunity for the reshuffle and a new layout of an industry. For MNCs, currently their focus on standards has far outweighed that on products. To achieve maximum economic benefits, they have gradually implemented their patent strategies and standard strategies. The formation of standards generally has two ways. First, the so-called "legal standards" are formulated by governmental departments or international organizations on the basis of trade-offs and compromises. However, as technological transformation accelerates, this simple method of technology development tends to be lagging or absent. Therefore, another way of formation, "factual standards", emerges. As technology becomes increasingly complex, the intensification of products also becomes increasingly higher. It is difficult to establish technical standards by carrying out R&D alone. It requires the participants of technical standards to cooperate, make contributions and carry out intensive management. By demonstrating the advantages of technical standard system as a group, they can implement a full package of external technological licensing, understand and obtain the future benefits of the standard system and licensing. Hence, they are willing to carry out technological cooperation.

It is easy to see from the above analysis that, through R&D internationalization and the development of strategic technology alliances between enterprises, MNCs have strengthened their monopoly in the supply of technologies. To cope with this situation, enterprises in China should strengthen international cooperation and strive to join the technological R&D activities led by MNCs to blend in with the global mainstream trend of technology development. Joining the R&D activities led by MNCs is a very good compromise for local enterprises of China that lack sufficient competitiveness for the moment. A good lesson from reform and opening-up over more than 20 years is that one should utilize the MNCs' investment in R&D internationalization to enhance enterprises' capabilities of technological absorption and innovation as well as improve the technological level of Chinese enterprises. They should accept the global strategic arrangements of MNCs and undertake processing for the transferred manufacturing industry to learn in cooperation and improve themselves. They should accept the primary technology development projects of MNCs (R&D adapted to local market) and enhance their capabilities through such cooperation. They should actively participate in the major scientific and technological R&D activities of MNCs, undertake part of the research and tasks of contributing resources. Even if they cannot decide the direction of development of advanced technologies, they should participate, give full play to their strengths, cooperate well with MNCs and gradually grasp the initiative. In this way, through continuous innovation, they should develop the core technologies with independent intellectual property rights, become a strategic partner that compete and cooperate with MNCs, and finally rely on indigenous innovation to break the monopoly of MNCs over core technologies at a higher level.

2. Breakthrough Technology Lock-in Using the Oligopolistic Reaction Strategy

As revealed by the analysis of the level of technology transfer and modes of technical cooperation between MNCs, the technologies of sole proprietorships are more advanced than those of joint ventures and the technical level of foreign holding enterprises among joint venture is higher than that of the non-holding ones. This is a problem about technical control and confidentiality of MNCs. In fact, the so-called technology spillover effect has long been included into the risk assessment of MNCs' technology transfer. MNCs adopt operation models that feature the trend of holding and sole proprietorship in China for their R&D institutions, which has allowed them to institutionally command the strategic height in technological control in China and to the greatest extent block the channels of technology diffusion to Chinese enterprises. Its real purpose is only to obtain China's relatively abundant yet cheap labor and even to plunder the original technological achievements of local enterprises and research institutes in China, thus better occupying the Chinese market and supporting global innovation network services.

In addition, the strategies of "division of labor of R&D chain" and "technical standards" of MNCs in China determine that the ultimate goal of local enterprises in China is indigenous innovation. Joining the global innovation alliance is only a necessary step toward indigenous innovation as well as a means and a method to improve China's technological capabilities. For the R&D institutions of MNCs in China, apart from some that are engaged in innovation-based R&D, most R&D activities are only targeted at the Chinese market, undertaking only promotion activities for adaptive, special technological and experimental products, i.e., the goal of R&D activities is to localize the products. This means that foreign-funded enterprises in China use the strategy of "division of labor of R&D chain" to control Chinese enterprises. In other words, basic and original research is carried out in home countries of MNCs while R&D institutions of foreign-funded enterprises in China mainly meet the needs of market development and occupation. The strategic goal is to allow MNCs to be located upstream in the industrial chain. Through detailed and specific division of labor for R&D, they monopolize and control core technologies, so that domestic enterprises of China are "enslaved" in the downstream section of industrial chain.

China's failed attempt at the "market-for-technology" strategy has fully demonstrated that wishful thinking can never help fulfill one's intentions. It is essential to promote indigenous innovation of local enterprises in China and improve the industrial technologies in China by driving MNCs to transfer more advanced technologies, while improving the strength of China's local enterprises in technological innovation through institutional settings and constraints with market mechanism. How oligopolistic reaction appears has been mentioned above. It is a kind of strategic response that MNCs make out of the need for mutual competition. Oligopolistic reaction only occurs between the entities engaged in mutual competition. However, since the fundamental purpose of their competition is to compete for resources and markets, if local enterprises of China can make full use of the special status of the market of host countries, it is also a possible strategic choice to introduce two or more strategic competitors and make them compete to adopt advanced technologies.

The two giants in the air separation industry, Linde from Germany and Air Liquide from France, competing to launch the most advanced technologies in China is the best example of breaking technology lock-in with oligopolistic reaction (see case study). The speed and level of Shanghai Volkswagen's acceleration of technology transfer after China's accession to the WTO also illustrates the effect of this strategy. Therefore, domestic enterprises with certain strength should be cultivated and/or a number of MNCs with similar strength in the same field should be introduced to replace "market for technology" with "competition for technology" and adopt the "law of oligopolistic reaction" to counter "technology lock-in strategy", so as to fully improve the quality of technology spillover effect of MNCs' R&D internationalization and promote the indigenous innovation of local enterprises in China.

3. Making Enterprises the Subjects of Technological Innovation

The role of Chinese enterprises as subjects in technological innovation is not prominent, their desire and motivation for innovation are insufficient, their personnel structure for R&D is unbalanced and their R&D funds are not ample. This chapter offers corresponding analyses from different angles with the foci of different chapters considered, but the most important problem is the mismatch between incentives for technological innovation of enterprises and the constraint mechanism.

Under the conditions of market economy, enterprises as the subjects of the market run the business on their own and take full responsibility for profits and losses. They select and determine innovative projects on their own and bear corresponding profits and losses, which is in line with the logic of technological innovation and business operations. However, for Chinese enterprises, technological innovation has always been a dilemma. In the 1990s, there was a popular saying among some enterprises that goes, "If you don't undertake technical transformation, there is no way out; if you do, you are playing with fire." This is a vivid portrayal of the dilemma. Therefore, it is necessary to understand the dynamic mechanism of technological innovation for enterprises under the conditions of market economy and to analyze how market demand works as a driving force and market competition as the pressure from the perspective of the relationship between enterprises and market environment. On the one hand, market demand is the original driving force of technological innovation for enterprises. To meet market demand with their own products and services, enterprises first must adapt to the changes in market demand, which not only directly affects the level of sales and revenue of products but also provides enterprises with new market opportunities and ideas for innovation. Enterprises should apply new technologies to meet the changing market demand through innovative activities and obtain the basis for their survival and development. On the other hand, market economy is also a kind of competitive economy where all competitors are trying to gain a competitive advantage. This requires enterprises to continuously develop new products, improve product quality, increase market share, refine equipment and production processes, and reduce product costs. As a result, enterprises are motivated to carry out technological innovation activities in a faster, better and more effective manner than their competitors, or they will be eliminated. Enterprises under competitive pressure will strive to transform the pressure from market competition into a continuous driving force for technological innovation for their own survival and development, constantly seek competitive advantages, and try to maintain such advantages for a long time.

Now that enterprises need to make technological innovation from the perspective of both market demand and market competition, why have Chinese enterprises not become the subjects of innovation? The reason is that there are problems with the incentive mechanism for technological innovation of enterprises. According to Schumpeter's innovation theory, technological innovation refers to entrepreneurs' new combination of factors of production. The essence of entrepreneurship lies in destroying the existing sources of advantages. This kind of creative destruction of entrepreneurs means that they are not satisfied with the status quo and are always ready to break the original balance through innovation and gain new advantages. Since innovation can create second-mover advantage and a certain degree of monopoly and lead to high returns, entrepreneurs have strong incentives to engage in development and innovation. In Western countries, enterprises are established according to the modern enterprise system and entrepreneurs are the initiators and organizers of technological innovation. They capture potential business opportunities in the market with a unique business vision, dare to take risks, choose and determine innovative projects and strive to maximize commercial profits by optimizing the allocation of resources. Successful innovation will then lead to improvement in business performance. This not only proves the vision and ability of the entrepreneurs but also satisfies their sense of accomplishment from self-fulfillment. In the modern enterprises and organizations, which feature the separation of ownership and control, the entrepreneur compensation system that is designed based on enterprise performance gives entrepreneurs material incentives for innovation. In China, the business operations of entrepreneurs in private enterprises are restricted by the imperfect market competition. Risks and benefits are asymmetrical and the enthusiasm for innovation is to some degree interfered. The state-owned enterprises feature completely deformed incentive mechanisms for entrepreneurs due to the unreasonable institutional design and distorted business culture.

Another issue is the constraint mechanism of corporate technological innovation. Enterprises face greater risks and uncertainties in technological innovation. As the subjects of technological innovation, enterprises must face such risks and uncertainties, and be constrained in all aspects, including resource capabilities. Firstly, there is the constraint mechanism for risks. Since enterprises have to bear enormous risks of technological innovation, while asking for corresponding returns on risk, i.e., to enjoy the economic benefits brought about by successful innovation, they must also build an effective risk constraint mechanism to control risks within an affordable range. Such a mechanism, while controlling risks, also brings about more cautious action plans to implement the innovative decisions of the enterprises. The managers of state-owned enterprises tend to pursue stability in terms of management, and without external intervention, they may voluntarily give up innovation. Secondly, there is the constraint mechanism for capital. Strongly constrained by corporate resource capabilities, technological innovation requires a large amount of capital investment, including not only capital investment necessary for R&D but also that for the purchase of test equipment, the introduction or training of technicians, and market development. Without sufficient capital investment, it is difficult for enterprises to guarantee the expected results of technological innovation. Therefore, corporate technological innovation activities are subject to the hard constraint from capital. China's private enterprises are still in their infancy, with poor capital strength and few financing channels. They are often forced to give up some innovation opportunities because of financial reasons.

Therefore, measures that help change the impetus for technological innovation of enterprises and transform enterprises into the subjects of technological innovation include designing a modern enterprise system, giving full play to the entrepreneurs' spirit of innovation, encouraging R&D personnel and general employees to participate in the technological innovation activities of the enterprises, broadening the financing channels of private enterprises for technological innovation, and ensuring a positive correlation between technological benefits and achievements of technological innovations.

4. Promoting the Construction of National Innovation System

Whether a country's technological innovation system is reasonable and effective should be analyzed in the following three aspects based on economics and system theory. First, it is necessary to investigate whether the allocation structure of technical resources is reasonable, i.e., the distribution of human resources, material resources, and information resources among the institutions and organizations that constitute the innovation system. Second, the ways in which the various elements in the technological innovation system are related to each other and interact with each other and the intensity of such connection and interaction also matter. The focus is the degree of their integration and synergy. Third, it is essential to examine the composition, function and operational mechanism of each subject of technological innovation behaviors (Zou, 2002). Through review, collation and summary of China's technological innovation system, it is not difficult to see that there are still some outstanding problems in China's technological innovation system.

First, the distribution of R&D personnel in China is unreasonable. At present, more than 50% of R&D personnel in China are found in universities and research institutes. R&D personnel in enterprises take up a lower proportion and tend to have lower academic degrees and professional titles than those in universities and research institutes. Developed countries not only have a large number of R&D personnel but also a large proportion of R&D personnel in enterprises, generally between 60% and 80%. This is because China's distribution system is unreasonable and people's perception of career choice is outdated. Many people think that those working at universities and research institutes have higher social status and more stable income and are respectable while those working in enterprises have poorer working conditions and lower payment. Hence, they try to find a job at universities and research institutes. Besides, the institutional changes over more than 30 years after the founding of the People's Republic of China have given rise to a fragmented technical management system. The rational flow and recombination of technical resources around economic construction are strictly restricted. The talent flow faces many hard constraints and too high personal costs, seriously hindering the optimal allocation of technical resources. Fortunately, with the economic benefits brought about by indigenous innovation of enterprises and the reform of national distribution system, a portion of R&D personnel has entered enterprises. However, structural changes still depend on the reform of the national scientific research management system while the updating of perception of career choice requires the efforts of several generations.

Second, Chinese enterprises fail to allocate sufficient funds for R&D. According to international standards, the intensity of corporate technological innovation is measured by the ratio of corporate input of R&D funds to corporate sales revenue. Chinese enterprises have a ratio far lower than those of developed countries and even many other developing countries. The development and experience of Western countries offer two obvious pieces of information. First, they maintain a higher level of technical expenditures. Second, in terms of the source and use of R&D funds, enterprises are the subjects of investment in the countries.

In addition, according to the *World Competitiveness Report* of IMD (2000 and 2001), China ranks 30th and 33rd in terms of international competitiveness in technology and 35th and 49th in terms of technological cooperation between enterprises, which is a sub-indicator. China lags behind all the developed countries. It ranks 30th in terms of adequacy of cooperative research among enterprises, universities and research institutes, which fully reflects its lack of synergy among the participants in the technological innovation system and the more serious structural defects of the system.

Hence, China must build and improve its technological innovation system in the following aspects. First, based on institutional innovation, China should build a macro-institutional environment suitable for the sustainable development of technological innovation, especially by strengthening the technological innovation service system for private enterprises and small and medium-sized enterprises, and form a modern enterprise system to boost entrepreneurs' spirit of innovation. Second, China should refine the technological innovation system framework with enterprises as the subjects, continue to optimize the allocation of technical resources, strengthen the cooperation and coordination among the various subjects in the technological innovation system and ensure the full play of their respective advantages and the integrity and smoothness of the technological innovation chain for cooperation between enterprises, universities and research institutes. Third, on the basis of structural adjustment, China should improve the micro-mechanisms of incentives and constraints within the various subjects of the technological innovation system, protect the enthusiasm of innovators, avoid short-term behavior of technological innovation, and more importantly, safeguard the motivations of front-line and general employees for technological innovation, as well as create a good atmosphere for all staff to innovate.

Chapter 8

Case Studies on the Air Separation Industry in China

It is generally believed that the FDI of the MNCs has a more positive than negative effect on host countries. However, in the recent years, economists are increasingly inclined to a middle ground. Both the positive and negative effects of the entry of MNCs are obvious. While naturally bringing technology spillovers, their entry also has crowding-out and substitution effects on local companies of host countries. The conclusion is relatively vague. There are few papers that have specific conclusions for different industries and have been dedicated to an in-depth discussion of the impact of the FDI of MNCs on a certain industry. In this book, we conduct an empirical analysis based on air separation, a major area of businesses of Linde Engineering (Hangzhou) Co., Ltd., where the authors work, and further study indigenous innovation as characterized by FDI of the MNCs as well as the relationship between technology transfer or introduction of such FDI in host countries and imitative innovation with absorption of technology spillovers as the main channel.

Gas separation and liquefaction units were primarily used to provide oxygen for the welding and cutting of metal as well as medical respiration. Later, as new technologies continue to emerge, it has been gradually applied to other industries in the secondary sector. In terms of metallurgy, it is used for basic oxygen steelmaking and blast furnace ironmaking. In the petrochemical industry, oxygen can be used to crack heavy oil to produce olefins and as raw material gas and fuel gas for synthesizing ammonia from coal powder gas. In industries such as textiles and electronics, nitrogen has been widely used as raw material gas, replacement gas and protective gas. In the industry of electric light source, argon, helium, neon, krypton, xenon and other gases are also used. In addition, products of gas separation and liquefaction units are widely applied in food industry, agriculture, animal husbandry and environmental protection. In the recent years, they have helped cure dozens of difficult diseases in the healthcare industry. The air separation industry is an industry born out of the design and manufacture of air separation units. In fact, apart from the use of cryogenic technology to liquefy and separate air, there are also methods, such as pressure swing adsorption and membrane separation. Since they separate a small amount of air and do not generate large economic aggregate, they are not covered in this book.

The air separation industry mainly includes two directions of businesses. The first direction is design, manufacture and installation of air separation units based on the requirements of the owners and delivery of complete sets of units. The main business feature is the general contracting of the project and the contractor is usually an engineering company. The other direction is the establishment of air separation units and delivery of the gases with various components to the user. The main business feature is the sale of gases and the provider is generally a gas company. Engineering companies belong to the typically technology-intensive industry. A complete set of air separation units consists of several complex systems. With the increasing capacity of air separation units (oxygen generators), many advanced technologies have emerged. Today, air separation technologies have developed to the level of the seventh generation. Many core technologies have been widely used, significantly improving the safety and reducing energy consumption of air separation units. Modern air separation units produce gaseous or liquid products of varying capacity and purity as well as oxygen and nitrogen with ultra-high purity. Therefore, in this book, we will carry out empirical study based on the technological innovation and progress of engineering companies engaged in building complete set of air separation units.

1. Oligopolistic Competition Situation of the Industry

The world air separation industry is dominated by the United States, Europe, Russia and Japan. The main enterprises include Linde from

Germany, Air Liquide (AL) from France, Air Products and Chemicals, Inc. (APCI) and Praxair, Inc. (Prax-air) from the United States, British Oxygen Company (BOC) from the United Kingdom, Messer from Germany, Cryogenmash from Russia, NIPPON SANSO K.K., Hitachi, Ltd. and Kobe Steel, Ltd. from Japan. At present, Linde, AL and APCI have set up R&D institutions in China. Among them, APCI has some defects in terms of employment and localization of technologies, with ordinary performance in the Chinese market. This section mainly introduces Linde and AL. China's manufacturing industry for air separation units began in the 1950s with the Hangzhou Oxygen Generator Factory. Through the construction of factories in the third-tier cities in the 1960s, a situation of tripartite confrontation among Hangzhou Hangyang Co., Ltd., which dominates, Kaifeng Air Separation Group and Sichuan Air Separation Equipment Group Co., Ltd. has emerged. In addition, there are manufacturers, such as Jiangxi Oxygen Plant Manufactory, Suzhou Oxygen Plant Co., Ltd. (SOPC) and Handan Oxygen Plant Manufactory.

The investment and income models demonstrated in Chapter 7 define the relevant factors of the R&D investment and income of MNCs. For better understanding, this section focuses on the basis for an owner's decision-making for investment on a set of air separation units (Sheng, 2005). The general situation is as follows. Back then, there were two available plans: domestic and foreign. The foreign plan represented by Linde was an investment of ¥250 million for annual production of units. It took 2 years to design, manufacture and install the plant, whose designed service life was 20 years. There were 12 operation and maintenance personnel, with an average annual salary of ¥50,000. The average annual overhaul time was 10 days and the average overhaul cost was ¥1.5 million. The main consumption after normal production was ¥192,240 per day and the sales income was ¥324,000 per day. This plan was characterized by leading technologies and high investment. The domestic plan represented by Hangyang was an investment of ¥200 million for annual production of the units. There were 20 operation and maintenance personnel, with average annual salary of ¥50,000. The average annual overhaul time was 20 days and the average overhaul cost was ¥2 million. The main consumption after normal production was ¥224,592 per day and the sales income was ¥324,000 per day. The features of this plan include feasible technologies and economical investment.

As required by the investing enterprise, the average return on capital was 8% and the enterprise income tax rate was 33%. The depreciation was calculated by the straight-line method and the residual value of the expired equipment was zero. On this basis, fixed assets investment and operating cash flow were calculated to figure out the NPVs of the two plans, which were ¥68.8 million for Linde's plan and ¥28.04 million for Hangyang's plan. Then their respective internal rates of return were calculated, namely, 11.41% for Linde's plan and 9.79% for Hangyang's plan. Therefore, considering the feasibility of project investment, both plans were feasible. Although the initial investment of Hangyang's plan was lower, the obvious advantages of Linde's technological solutions could greatly reduce the actual operating costs of the owner. From the perspective of the life cycle of air separation unit, Linde's plan was better. Based on the principle of the highest cost effectiveness, the owner finally chose Linde's plan.

Such a choice is the result of the rational decision of the owner (consumer) and also provides a basis for decision-making by large MNCs like Linde to set up corresponding R&D centers in China and invest an appropriate level of technology. Therefore, the advanced technologies of MNCs feature high initial investment, low operating costs, stable product quality, as well as easy and controllable operation. The main reason for the high initial investment is not entirely the hardware cost of the plant, but more the excess profit that MNCs make by monopolizing technologies. The Chinese government advocates a number of economic policies, such as circular economy and energy conservation, which have provided unprecedented opportunities for MNCs to take up more market space. In the recent years, the technological development of the air separation industry in the Chinese market has shown obvious oligarchic characteristics, which are mainly as follows.

There are two major foreign manufacturers:

(1) Linde was founded in June 1879. On May 29, 1895, Linde made the first high-pressure air liquefaction plant with a capacity of 3 L/h in the world based on the Joule−Thomson effect, put it into industrial production and established the Linde liquefaction cycle with throttling process. In 1903, Dr. Linde designed the first industrial 10 m³/h oxygen generator in the world, using a high-pressure single-tower throttling process. Oxygen generator was thus born. From the 1960s to the 1990s, Linde produced air separation units with a capacity ranging from 11,000 m³/h to 70,000 m³/h. In the current global manufacturing industry for air separation units, Linde, AL and APCI have

been competing with each other, which brings up the scale of air separation units. The output has increased from 1,000 t/d to 3,500 t/d. The largest nitrogen generator manufactured by Linde is the one with a capacity of 3,350,000 m^3/h installed in Mexico, which is equivalent to the air processing volume of 3,500 t/d oxygen generator and 10,000 t/d nitrogen generator. Hence, Linde also claims to have produced the largest air separation unit in the world. In 1995, Linde and Bingshan Group jointly established Linde Process Plants Co., Ltd. In May 2002, Hangzhou Office (Engineering and Sales Center) was established. In 2005, it was changed to Linde Engineering (Hangzhou) Co., Ltd. and controlled by Linde. In this way, the design, manufacturing, project execution and procurement of Linde's air separation units in China were gradually localized.

(2) AL was built in 1902, which was also the year when Claude invented the piston expander and established the "Claude liquefaction cycle". In 1910, the first medium-pressure oxygen generator with an expander and a capacity of 10 m^3/h was made. Thus, the production of air separation units started. In 1957, the air separation unit with a capacity of 10,750 m^3/h was built, followed by large ones with a capacity ranging from 16,000 m^3/h to 74,000 m^3/h. The largest air separation unit designed and manufactured by AL is the one with a capacity of 3,350 t/d (approx. 10,366 m^3/h) for the order from Sasol Secunda in South Africa. This indicates that AL has the ability to provide equipment and services to users around the world who have a huge demand for oxygen and nitrogen. At present, the largest air separation unit in operation in the world has the capacity of 3,200 t/d (93,440 m^3/h) and was manufactured by AL. It set up a subsidiary controlled by it in cooperation with Hangyang in Hangzhou in January 1995. The subsidiary has become one of the five pillars of AL's global engineering design and manufacturing, and has a complete set of technologies that are transferred from AL and constantly updated. Its products represent today's international trends of technological development and the applications of the latest technological achievements in this field.

There are three major Chinese manufacturers:

(1) Hangyang was established in 1950. In the late 1950s, it successfully developed the first generation of high-/low-pressure oxygen

generators with aluminum strip regenerators. Since then, its technologies have maintained rapid progress. By the early 1990s, the sixth generation of air separation unit with structured packing, full distillation and hydrogen-free argon production technology had been successfully developed. In 2001, it developed and designed the air separation unit with a capacity of 30,000 m^3/h, which filled in the gap for such capacity in China. In June 2004, the complete set of air separation products with a capacity of 50,000 m^3/h was successfully developed. Currently, domestic large-scale air separation products have reached a new level. Hangyang's main products are air separation units and related products of various specifications. It is the leading manufacturer of air separation units in China. With the deepening of R&D, the large and medium-sized air separation units and processes designed and produced by Hangyang tend to be diversified and the single-unit capacity tends to be large. About 20 sets of air separation units with a capacity of over 30,000 m^3/h have been sold. The first set of large-scale air separation units with a capacity of 50,000 m^3/h in China designed and manufactured by the company was successfully launched in June 2004. At present, Hangyang has become one of the major manufacturers of air separation units in the world, with large and medium air separation units exported to more than 30 countries.

(2) Sichuan Air Separation, formerly Sichuan Air Separation Equipment Factory, was founded in 1962. It is mainly engaged in the design, manufacture, sale and installation of large, medium and small air separation units and related products. It is one of the major designing and manufacturing bases for large air separation units in China. In terms of large-scale air separation units, nearly 30 sets whose capacity ranges from 10,000 m^3/h to 20,000 m^3/h have been successfully put into operation. The company has fully mastered the key technologies for large air separation units and ranks among the best in terms of technological level in China. Through cooperation with many domestic universities in research and many famous international companies in production, Sichuan Air Separation has digested and absorbed international advanced technology and actively made innovation, which has greatly improved the technological level of products. It takes the lead in China in terms of overall technological level. The capacity of its air separation units has reached 30,000 m^3/h and that of the largest unit it produced in cooperation with foreign enterprises is 40,000 m^3/h.

(3) Founded in 1965, Kaifeng Air Separation is a large-scale backbone enterprise in China's air separation unit industry. Since the "Seventh Five-Year Plan", it has invested a lot of energy in technological transformation and evolved from an enterprise positioned at producing medium air separation units in the period of planned economy to the current one specializing in design and manufacture of complete sets of largest air separation units in China. It took the lead to put the localized 40,000 m³/h large air separation unit to operation in 2004. In 2005, it began the design of the air separation unit with a capacity of 53,000 m³/h for Henan Yongcheng Coal and Electricity Group. Since its founding and operation, Kaifeng Air Separation has provided nearly 500 complete sets of large and medium air separation units for many industries in the secondary sector in China. Since the 1990s, more than 40 sets of products have been exported. At present, the air separation units designed, manufactured and installed by the company are among the best in the industry in China in terms of product specifications, technological level and quantity of complete sets. Table 1 shows the top five manufacturers in the air separation industry ranked by major economic indicators in 2008.

2008 marks a period of recovery for the gas separation unit industry after the financial crisis. The whole industry maintained the momentum of rapid growth, with a year-on-year growth of industrial output value at 39%. For example, the total industrial output value of Hangyang and Sichuan Air Separation for the period from January to September reached the level of 2007; the growth rate of the industrial output value of Kaifeng Air Separation Group also exceeded that of the same period in 2007. By the end of 2008, the assets of the air separation industry totaled ¥16.79 billion and the fixed assets totaled ¥1.67 billion. The number of employees was 14,267, including 2,525 engineers and technicians. The air separation industry adhered to technological innovation, strengthened technical management, and further improved production capacity and economic benefits. In 2008, the whole industry produced a total of 289 sets of air separation units, with a total oxygen production capacity up to 3,063,600 m³/h. This is mainly because the orders for large air separation units increased. The sales revenue from products reached ¥10,245,320,000, an increase of 14.58% year on year.

To facilitate the analysis of main competitors' market share in China's air separation industry and the contribution of the technology factor to the

Table 1.   Top Five Manufacturers in the Air Separation Industry Ranked by Major Economic Indicators in 2008*

Indicator	1st	2nd	3rd	4th	5th
Gross industrial output value at current price (¥10,000)	Hangyang 436,838	Sichuan Air Separation 243,048	Kaifeng Air Separation 77,518	Linde 77,330	Hebei Air Separation 58,373
Industry value added (¥10,000)	Hangyang 117,744	Sichuan Air Separation 87,672	Linde 23,350	Hebei Air Separation 12,566	Kaifeng Huanghe Air Separation 12,566
Sales revenue from products (¥10,000)	Hangyang 455,831	Sichuan Air Separation 202,340	Linde 77,330	Kaifeng Air Separation 74,416	Hebei Air Separation 58,038
Total profits and tax	Hangyang 90,711	Sichuan Air Separation 26,466	Linde 18,031	Kaifeng Air Separation 4,497	KFDJ Air Separation 2,647
Quantity of air separation unit (set)	Hangyang 62	SOPC 46	Sichuan Air Separation 42	Handan Oxygen Plant Manufactory 42	Kaifeng Air Separation 24
Quantity of air separation units with a capacity above 30,000 m³/h	Hangyang 36	Sichuan Air Separation 30	Kaifeng Air Separation 18	Hebei Air Separation 18	Kaifeng Huanghe Air Separation 12
Economic benefits composite index (%)	Hangyang 288.11	Sichuan Air Separation 286.76	Kaifeng Huanghe Air Separation 281.00	KFDJ Air Separation 228.82	Hebei Air Separation 199.85

Note: *Industrial Statistical Yearbook on Gas Separation Equipment. Gas Separation Equipment Branch of China General Machinery Industry Association, 2009-04-26.

market share, we specifically rank the top five manufacturers based on the main economic indicators of the air separation industry in 2008 and the results are shown in Table 1. The table conveys two very important pieces of information: the market shares of manufacturers and the product structure and technological content of the manufacturers.[1] Obviously, the market share and sales revenue of local enterprises in China are in an absolute dominant position. However, their proportion has dropped significantly in terms of the quantity of large-scale air separation units with a capacity above 30,000 m^3/h. None of them has produced extra-large air separation units with a capacity over 60,000 m^3/h. As for economic benefits, MNCs are far ahead of local enterprises of China, the reason for which is very simple. MNCs obtain excess monopoly profits by virtue of leading core technologies.

2. Trail of the Technological Development of the Industry

Since the first 10 m^3/h oxygen generator was produced by Linde in 1903, MNCs have been developing air separation units and technologies for more than 100 years. Yet, the industry in China has a history of only 50 years, which is one of the objective reasons why China lags behind in terms of technology in this industry. However, on the other hand, the technological gap between China and MNCs is shrinking. The gap is about 15 years in terms of the capacity of air separation units and less than 10 years as to how advanced the processes and procedures of air separation units are. The narrowing of the gap is the result of technology introduction by local enterprises in China and FDI of the MNCs. The trail of technological development of the air separation industry in China is as follows.[2]

Before the founding of People's Republic of China, there was no manufacturing industry for gas separation and liquefaction unit in China. There is little application of this kind of equipment. From the first 15 m^3/h (oxygen) air separation unit purchased from Japan in 1934 to the eve of the founding of People's Republic of China in 1949, only a few coastal

---

[1] Industrial Statistical Yearbook on Gas Separation Equipment. Gas Separation Equipment Branch of China General Machinery Industry Association, 2009-04-26.
[2] http://www.cngspw.com/jishu/ViewThesis.asp?DocID=TY2004M4D15 H173744,2004-4-15.

cities such as Shanghai and Qingdao used imported air separation units, which totaled less than 100 sets. The oxygen production capacity of a set fell between 10 m^3/h and 200 m^3/h.

The first set of air separation units with a capacity of 30 m^3/h (oxygen) was developed in the early 1950s. Harbin Oxygen Generator Factory hired five Russian engineers and succeeded in trial production of two sets of (oxygen) air separation units with a capacity of 30 m^3/h at the end of 1953. Starting in August 1953, Hangyang analyzed the equipment imported from the Soviet Union and obtained some key drawings of the equipment. Then, in January 1956, it started mass production of (oxygen) air separation units with a capacity of 30 m^3/h.

In 1956, Hangyang began to produce medium air separation units with high- and low-pressure processes. Hangyang analyzed the air separation units with a capacity of 30 m^3/h (oxygen) imported from the Soviet Union, went to the Soviet Union for field study and hired Soviet experts. Eventually, it succeeded in trial production of the first set of (oxygen) air separation units with a capacity of 3,350 m^3/h at the end of April 1958. At the same time, it completed the transition from high pressure process to medium pressure process with reference to the (oxygen) medium pressure process with a capacity of 50 m^3/h in Germany. In September 1958, Hangyang established the first "Oxygen Production and Refrigeration Equipment Research Center". In April 1960, Hangzhou Oxygen Generator Research Institute was established to undertake research and design of rare gas extraction and liquefaction equipment, and serve as an industrial center of technological information. In July 1961, it launched a professional technical publication. The period from 1956 to 1969 marked the formation of China's gas separation unit manufacturing system. The gas separation and liquefaction unit manufacturing system of China emerged, dominated by the eight factories and two institutes in the industry.

From the early 1960s to the end of the 1970s, China's air separation industry saw the development of a large number of key technologies and units, such as rare gas extraction unit and gas liquefaction unit, liquid oxygen and liquid nitrogen unit, hydrogen liquefaction unit, helium liquefaction unit, neon liquefaction unit, cryogenic storage tanks and counter-turbine expanders. It is particularly worth mentioning that the company learned a painful lesson from the experimental research on the plate-fin heat exchanger. Everything turned out to be useless because its absorption of key technologies was insufficient and the project was carried out in a rush. The main technical personnel in China had discussions and formulated general technical requirements for the plate-fin heat exchanger and seven regulations on the processes for manufacturing of parts, precision, cleaning, final assembly, brazing, testing, etc. In the end, the plate-fin heat exchanger was completed under the policy of "self-reliance", laying the foundation for large and medium full-plate air separation units.

In 1978, the relevant state departments began to seek new ways to fundamentally solve the problems concerning technological level and product quality of large air separation units. By combining technology and trade, they introduced technologies for two types of 10,000 m³/h air separation units from Linde, which provided technical data about design, calculation, manufacturing, quality control and testing, and cooperated with them on production. During digestion and absorption of the introduced technologies, Hangyang tested and compared the original product series and, through comparative analysis of new and old design methods, found out the cruxes of the poor performance, low index and unreasonable design of domestic large air separation equipment. At the same time, for some key structural designs of the imported technologies, research projects were carried out for test and verification so as to master key technologies. Through absorption and digestion of the introduced technologies, the technological level of domestic large air separation units was significantly improved.

From 1986 to 1990, Hangyang studied the calculation of the then latest international air separation process with booster expansion and mastered the difficult technologies, such as process calculation methods, parameter selection, and efficiency of converting work of expansion into that of compression. It cooperated with Xi'an Jiaotong University to complete the key technologies of design and manufacture of large units and successfully applied them to the first 6,000 m³/h air separation unit with booster expansion process in China, realizing the technological upgrading of the whole series of air separation units in China. Since 1992, manufacturers such as Hangyang and Kaifeng Air Separation have cooperated with universities such as Tianjin University to complete the development of structured packing, gaining experience in developing new generation of air separation unit using structured packing upper tower and full distillation technology for argon production.

From 1993 to 1997, China completed the development and production of its first large liquid oxygen internal compressed air separation unit and the research and development of the first set of air separation units with full distillation for argon production. By the end of 2000, on the basis of independent development of air separation unit with molecular sieve for purification and booster expansion process, Hangyang had mainly undertaken process design, calculation and control, conducted analysis and research based on experience in both testing and debugging, and proposed feasible procedure, i.e., it established a calculation model to develop software on its own. In terms of control, stable upper tower conditions and extraction of argon fraction had been ensured to improve the extraction rate of argon. In this way, it completed the development of a new generation of air separation unit. The technical performance of large and medium air separation unit was much better than the fifth generation in terms of technical performance, marking a new leap of the new generation of large and medium air separation unit technologies in China.

On September 24, 2001, Hangyang signed a contract to supply Shanghai Baosteel with the first set of 30,000 m³/h air separation units, the first of its kind designed and manufactured by Hangyang independently. In terms of design, booster expansion for refrigeration and two advanced air separation technologies, namely, structured packing upper tower and full distillation for argon production, were used. The whole set of units came with reliable modular design software and advanced industrial control system for control of the whole set to integrate central, machine side and local control. This allows effective monitoring of the production process of a complete set of air separation units. The complete set was successfully put into use in 2002 and its output and purity of oxygen, nitrogen and argon products all met the design index. It even surpassed products of foreign companies in terms of the smooth level of debugging. All these mark that the systematicness and stability in design and production of extra-large air separation unit in China are close to the world's advanced level.

In 2008, Hangyang increased its investment in technology development. Progress was made in more than 90 technology improvement projects, with 12 new products and technologies developed and produced. Hangyang has been granted four patents and its applications for nine invention patents have entered the substantial approval procedure. It cooperated with Zhejiang University on the research and production of fin test bench, which has been delivered and put into use. The company also made great progress in information-based construction. The official versions of computer operating systems and office software have been used. The Hangyang PLM project, sales management project, information transmission for official documents, corporate system for protecting intellectual property rights and OA system have been preliminarily built.

On October 25, 2008, the 20,000 m³/h air separation unit that Hanging supplied to Nanjing Iron and Steel Co., Ltd. passed user assessment. The unit came with the advanced control technology of automatic variable load developed by Hangyang and Zhejiang University. The technology allows variable load operation of the product oxygen in the range from 15,000 m³/h to 20,000 m³/h, filling in the blank of the domestic air separation industry in automatic variable load technology.

3. Inspirations for Indigenous Innovation of the Industry

Today, in the air separation industry, the world is dominated by three major gas companies, namely, Linde, AL, and APCI, all of which have created ultra-large air separation units that demonstrate their advanced technologies and great strength. AL and Linde also each formed engineering companies with majority shareholding with local Chinese companies to undertake Chinese projects in 1995. They are responsible for localization activities for design, manufacturing, project execution and procurement, creating a close link between R&D localization and market competition activities. APCI established a wholly owned subsidiary in Shanghai in 1998. Although it had problems in integrating with local enterprises, its business has been developing well. Praxair, BOC, and MG are also trying to keep up. They do not reduce investment activities in China and have become the "three pillars" in the second tier in terms of expansion and competition. Air separation units are advancing rapidly in the continuous technological innovation of these MNCs, setting the trends of development of air separation in the world.

The 30,000 m³/h air separation unit that Hangyang provided to Baosteel in 2002 has successfully generated oxygen, marking the successful localization of large air separation unit with a capacity of 30,000 m³/h. Now, Hangyang, Sichuan Air Separation and Kaifeng Air Separation have begun to work on the development of extra-large air separation unit with a capacity of 60,000 m³/h. Technically, advanced technologies, such as molecular sieve for purification, booster expansion technology, structured packing tower technology, and hydrogen-free argon production technology, have been gradually adopted. The air separation process has gone through the fourth and fifth generations. Now the sixth generation has been widely applied. Technologies of air separation unit have developed rapidly. Though there is still a certain gap between them and the advanced foreign technologies, they basically meet the requirements of the domestic air separation market and have been well received by the users. The domestic air separation units are quite competitive in the domestic market.

As the trail of technological development reveals, the technologies of China's air separation industry have undergone several stages, including technological introduction, original innovation, digestion, absorption and reinnovation, as well as tracking of international advanced levels. Technological introduction was mainly realized by learning from the companies and research institutes of countries represented by the former Soviet Union in the initial stage and by studying drawings and analyzing prototypes. Due to the political background, the Soviet Union was completely open in outputting technologies to China, which has played a vital role in the building of China's air separation technology system and the training of technicians. Original innovation mainly took place from the 1960s to the 1970s, when China was in a relatively closed environment. Under the planning of the First Mechanical Bureau, the national technicians shared out work and tackled the difficulties for a series of major equipment and built China's air separation unit manufacturing system, making important contributions to China's economic development. However, at that time, the technological level of China's air separation units was lower; its gap with the world's advanced level continued to widen. From the late 1970s to the mid-1990s, enterprises like Hangyang started to introduce air separation technology for a capacity of 10,000 m^3/h from Linde and cooperated with it on production. Through digestion, absorption and reinnovation, there was substantial progress in the technological level of large domestic air separation units. From then on, basic research and application development became equally important.

Since the implementation of R&D internationalization strategy by MNCs in the air separation industry in 1995, i.e., since Linde and AL set up subsidiaries in China, local enterprises in China have had the opportunities for close learning and observation. They have been able to grasp the development trends of foreign technologies in time and, through connection with these companies and the upstream and downstream domestic partners, indirectly support the improvement of China's air separation level. In this way, China's industrial technologies for air separation have officially entered the stage of tracking the international advanced level. This has the following manifestations. The stock of human capital in the air separation industry has changed. The proportion of front-line employees, engineers, technicians and managers has changed drastically. The structure of human capital has been optimized, with a significant increase in the number of technical staff. By the end of 2008, there had been 2,525 technological professionals in the industry. The distribution of technicians of major local enterprises is shown in Figure 1. Local enterprises have been stimulated to make investment in S&T research and the contribution rate of technology to economic benefits of enterprises has increased year on year. In 2008, average contribution rate of S&T funds of the major local enterprises to the output value of new products was 6.3%. The labor productivity of all employees also continuously improved. The composite index of industrial efficiency grew rapidly year on year, as shown in Figure 2. Industrial competition intensified. Some small and medium-sized enterprises whose comprehensive strength was weak and which was insolvent were restructured, merged and reorganized to optimize the allocation of resources within the industry. The domestic supporting capabilities for air separation units were greatly enhanced, and the localization rate of key equipment was increasing steadily. The ratio of domestic supply of static equipment, pipelines and valves as well as instruments

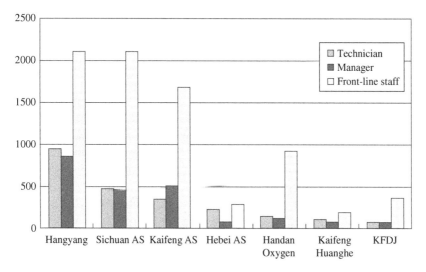

Figure 1.   Stock of Human Capital of Major Local Enterprises in Air Separation Industry in China at the End of 2008 (Person).

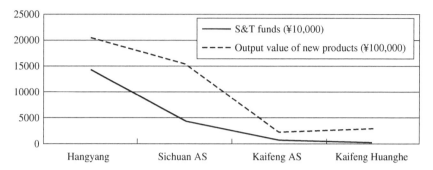

Figure 2.   Relationship Between S&T Funds and Output Value of New Products of Major Local Enterprises in Air Separation Industry in China in 2008.

greatly increased. Even the three major kinds of units that used to be completely imported were produced by some of the domestic backbone enterprises or supplied jointly by domestic and foreign companies. This effectively reduced the costs for domestic users and improved the cost performance of air separation units.[3]

Through further and objective analysis of the effects of the R&D internationalization of the MNCs on China's indigenous innovation, we summarize several related factors that influence the effects as follows:

(1) **Ways of investment and establishment of R&D institutions:** Based on the different ways of investment and establishment, R&D institutions fall into three categories, with varied influence on the technologies of host countries. The first category is the new wholly owned R&D institutions that are directly under the leadership of the headquarters in terms of research funds, staffing, research direction and research focus. They are independent. As the extension of R&D work of the headquarters, they are mainly designed to compete for the technological resources of host countries. The personnel tend to flow in one direction. The absolute amount of technological spillover is very limited. APCI's engineering companies in China belong to this category. The second category is the joint venture or R&D institutions founded upon merger. They are set up jointly or upon merger through reintegration based on the original technological strength,

---

[3] Industrial Statistical Yearbook on Gas Separation Equipment (2006). Gas Separation Equipment Branch of China General Machinery Industry Association, 2006-05-16.

with consideration over the original technological innovation capacity as well as the global production layout and the development of local market. After the joint venture or merger, through the technology input from the MNCs' headquarters, the original hard and soft technologies are often improved. Linde and AL fall into this category, as evidenced by their commitment to promoting the Chinese version of technologies. The third category consists of some R&D institutions that are established through technology alliance. Some MNCs establish strategic alliances with local efficiency via contract for the purpose of "supplementing resources", "avoiding war" and "increasing cooperation opportunities".[4] Such alliances include knowledge alliances and product alliances. The former aims to learn and create knowledge and to improve the core technological capabilities of enterprises, with strategic influence on the technological innovation of local enterprises. The latter, whose major target is productive technology, mainly develops technologies to support the production of host countries. This directly and rapidly improves the production efficiency of enterprises of host countries. The technical cooperation between Messer (MG) and Hangyang belongs to this category. Figure 3 shows the link between the methods of investment and establishment of R&D institutions and their technological effect. If the investment method leans towards cooperation, the average level of technical effect will be higher.

(2) **Technological level of local enterprises:** This is directly related to technological spillovers. If the technological level of local enterprises lags behind and the quality of manpower is lower, the technological competition in the market will not be very fierce. MNCs tend to transfer some products with low technological content to take advantage of the low cost of the host countries. On the contrary, if the local technological level is higher, MNCs tend to transfer technologies with higher value-add to make use of local advantages in technology and talent. In China, due to the existence of a number of domestic enterprises with strong technological strength, such as Hangyang, Sichuan Air Separation and Kaifeng Air Separation, as well as the needs of strategic competition between Linde and AL, the domestic competition is fierce, forcing these two MNCs to transfer and adopt new

---

[4] Hasegawa, S. Qtd. from Li, A. F. *R&D Internationalization of Multinational Corporations* (*1st Edition*). Beijing: People's Publishing House, 2004.

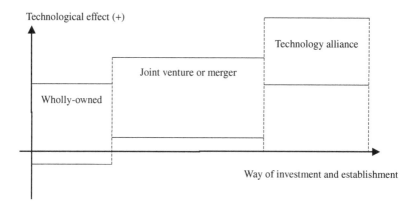

Figure 3.   Box Plot of R&D Institutions and Technological Effects.

technologies of their parent companies to improve their market competitiveness. This leads to more apparent technology spillover effects.
In the fierce market competition, local enterprises in China have
gradually developed stronger learning and imitative capabilities. They
are able to effectively utilize their own advantages to cope with competition from MNCs and systematically absorb the spillover technologies of MNCs. A virtuous circle is then formed during their
competition with MNCs, which further strengthens their technological advantages. Other domestic air separation manufacturers, such
as Harbin Oxygen Generator Factory and Handan Oxygen Plant
Manufactory, do not have enough technical support to absorb and
imitate the technologies of MNCs due to the large gap between them
and the MNCs, which results in slow development of technology and
continuous decline in market share. Therefore, the impact of technological gap on technology spillover effects is two-sided. Within a
certain scope, a larger technological gap means more opportunities for
learning and imitation, which is conducive to the absorption of technology spillovers. However, when a certain degree is reached, technology spillover effects will decline because the technology recipients
do not have enough technological capabilities to absorb them. There
is an inflection point in the middle, as shown in Figure 4.

(3) **Stock and flow of human capital:** Previous studies have shown that
the stock of human capital in the host countries is a key factor affecting the absorption of technology spillover effects. This is a kind
of static analysis. In fact, the stock of human capital is constantly

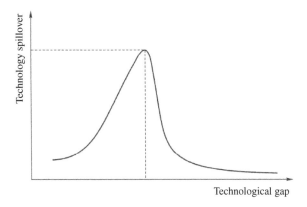

Figure 4.   Relationship Between Technological Gap and Technology Spillover Effect.

changing. For the air separation industry, the most important human capital is engineers, including personnel for technological design, project execution, project procurement and sale. First and foremost, they have received professional education in the field of engineering technologies at universities. This is a prerequisite for their becoming engineers. Then they embark on corresponding research positions and professional directions to engage in related practice and training sessions. With certain research experience and qualifications, they become engineers in the true sense. Therefore, after MNCs establish R&D institutions in China, they create jobs and there are new increments of human capital. As for the existing stock of human capital, when MNCs set up R&D institutions, they first of all absorb the employees of local companies with better salary and treatment, especially experienced engineers. Leaders in many positions of Linde and AL come from Hangyang. In this sense, they cause the one-way flow of talent. However, not all the best professionals have been attracted for two reasons. First, MNCs attach more importance to interdisciplinary talent. They need higher expertise as well as work skills involving English and computer. Given the Chinese education system, those who have greater business proficiency may not be suitable for MNCs. On the other hand, although there is a gap in salary between domestic enterprises and MNCs, a smaller portion of young and middle-aged people join MNCs considering the comprehensive welfare (including housing and equity) as well as people's pursuit of stability and security. After nearly 10 years of talent flow, the number of technicians that have left Hangyang for Linde and AL totals about 80, accounting for only 5% of the total number of employees in Hangyang. Thus, the impact on Hangyang is still very limited. In contrast, those who have left Hangyang continue to contribute to its current development through technical exchanges in the same industry as well as technical support when they purchase products from Hangyang's subsidiaries. Therefore, the R&D investment of MNCs does not alone promote the economic growth of the host countries; the combination of investment and human capital of the host countries is also a booster. China's air separation units evolved from the medium ones with a capacity ranging from 5,000 m³/h to 10,000 m³/h in the mid-1990s to the large ones with a capacity of 30,000 m³/h. Besides, technological revolution of full distillation for the extraction of pure argon has been completed. This benefits to a great extent from the most advanced technologies that Linde and AL are constantly bringing to China.

(4) **Degree of integration between MNCs and local production systems:** This is mainly reflected in the vertical link of the MNCs. In order to reduce production costs and speed up delivery, MNCs maximize the localization of procurement, which prompts them to provide training sessions and guidance to local supporting enterprises on technical standards, procurement conditions and quality control. In this way, productivity and the production level are improved. Over the past decade, China's local enterprises have mastered a complete set of key and core technologies for many large air separation units. The manufacturing of major components, such as static equipment, valves, steel structures and pipes, has reached the world's advanced level. The design and development of products have been improved to a certain degree. The localization rate of key equipment, which has begun to be exported to developed countries such as Europe and the United States, has surged to over 80%. Besides, the original foreign suppliers of MNCs, such as ABB and SIEMENS, followed the development strategy of MNCs and set up factories in China. They increased their market share in China through training and invitation to observation and learning. These supporting enterprises also supply high-quality raw materials and parts to local enterprises in China, improving the level of support for China's air separation industry and bringing strong stimulation to the technological progress of the host countries from the two angles. At present, in the air separation

industry, except the three major kinds of units that are usually imported directly from abroad, other equipment, instrumentation and pipelines are purchased from local joint ventures and pure local enterprises, which has generated obvious technology spillover effects on local enterprises.

Nowadays, air separation industry has developed rapidly in China, and local enterprises enjoy strong momentum of technological development. Domestic enterprises represented by Hangyang have successfully solved problems regarding manufacturing and design technologies of key equipment when developing oversize air separation units, effectively safeguarding their development and promotion. However, the localization of large air separation units should be understood as the domestic enterprises' grasp of a complete set of core and key technologies for large-scale air separation units, rather than mere scale and capacity. For domestic enterprises, localization is a matter of enhancement of comprehensive technical strength, and requires indigenous innovation capacity.

4. Puzzle of Indigenous Innovation of the Industry

Innovation is viewed as the soul of a nation and an inexhaustible driving force for a country to thrive. The improvement of indigenous innovation capacity is an essential strategic option for China in the time of economic transformation and the key to realizing fundamental transformation of economic growth mode. At the 2005 Annual Meeting of the China Association for Science and Technology, Chen Zhili stated that indigenous innovation was an inevitable strategic option to solve major bottleneck problems of national economy, such as unreasonable structure and extensive growth mode. According to him, indigenous innovation is the strategic arrangement for addressing the problem that key technologies are controlled by other countries and a major deployment to solve the problem of improving national competitiveness.[5] In the context of economic globalization, the possibility that countries use external technological resources has greatly increased and China's rapid economic

---

[5]Summary of Chen Zhili's speech at the 2005 Annual Conference of the Chinese Association of Science and Technology. http://scitech.people.com.cn/GB/1056/3631349. html.

development seems to make it possible for Chinese enterprises to make technological progress. However, over the past 28 years of reform and opening-up, there has been no substantial improvement in technological level internationally, and the rankings of international competitiveness have also continued to decline in the recent years. The main reason is that China's technological dependence on MNCs has deepened in the process of opening up. The environment for technological innovation has been destroyed, and motivation for indigenous innovation is woefully inadequate.

4.1. *Technology investment and management*

Technology investment is a requirement and a basic guarantee for techno-logical progress. Increasing technology investment is a significant means to improve technological innovation capacity, promote social and eco-nomic development and enhance overall national strength. From the per-spective of nation and government, China always has a problem of inadequate technology investment. Historically, the highest proportion of technology investment to GDP was 2.32% in 1960, and then the figure declined year on year to only 0.69% in 1998. It had rebounded since 2000 and increased to 1.23% in 2004, which has doubled compared with the lowest point. However, it still did not reach 1.5% as set in the relevant regulations in China. At present, R&D funds of many developed countries have accounted for more than 2% of GDP. The figure is close to 3% for Japan, South Korea and the United States, and as high as 4.7% for Israel. In light of this, compared with great and major powers of technology around the world, China is facing a shortage of technology investment. At the level of enterprise, there are greater differences. For many years, Chinese large and medium-sized industrial enterprises have had the inten-sity of R&D fund investment at around 0.7%, and private and individual enterprises have a lower proportion, while investment by enterprises in major developed countries has fallen between 2.5% and 4%.

Meanwhile, China's insufficient indigenous innovation also lies in simple and rigid management and evaluation system of technology invest-ment. For a long time, China has a problem of structurally deformed consumption of technology investment, and there have always been two typical problems in China's technology and business circles, i.e., focusing on introduction and hardware and overlooking digestion, absorption and software. For example, China's ratio of technology introduction to digestion and absorption of fund investment is 1:0.08 in 2002, while that of South Korea and Japan is between 1:5 and 1:8 during similar development period. The experience of South Korea and Japan is to introduce key equipment and technology, and truly realize that technology introduction is the premise, imitative innovation is the purpose, and indigenous innovation is the aim. However, since Chinese government lacks policy guidance on technology introduction, there are no special funds for digesting and absorbing introduced technology, and departments responsible for technology introduction are more likely to consider short-term behaviors and tend to introduce complete sets of technology. In the short run, complete sets of equipment and technology mean better performance, safer operation and more convenient maintenance. However, this greatly reduces opportunities for Chinese enterprises to learn and make improvement, and leads to heavy technological path dependence. The mismatch between technology introduction and digestion and absorption leads some industries into China to a strange circle of "introduction, backwardness, reintroduction, and backwardness again", restricting technological reinnovation and greatly reducing the effect of the limited technology investment. Currently, the degree of China's foreign technology dependence is hovering at 50%, while that of the United States and Japan is only about 5%. The degree of self-sufficiency of key technology is low. Of equipment investment that accounts for 40% of fixed assets investment, more than 60% needs to be imported, and key equipment with high technological content is basically dependent on imports.[6]

In 1999, China began to reform technology system, and research institutes generally started to implement internal management systems, such as president (director) responsibility system and appointment system for professional and technical posts. Employment system and contract system began to be the basic employment mode for research institutes. Technical staff had business assessment biennially or triennially. Employment would not be continued unless employers passed the evaluation. They might still lose their jobs if they failed to obtain scientific research achievements. The problem is that scientific research has its own laws, and research achievements do not come evenly like tap water. Current evaluation-based employment system forces technical workers to lower their standards and publish low-level articles. Meanwhile China's technology evaluation system is very simple and rigid. For a long time, many

---

[6]Nine Puzzles Challenging Innovations. *Qianjiang Evening News*, 2006-01-09(A15).

entities take the quantity of papers as the major evaluation indicator, resulting in blind pursuit of quantity as well as papers with disappointing quality. Between 1993 and 2003, there is no Chinese scholar among the top 20 authors whose SCI papers are most cited in all disciplines around the world and there are only two Chinese scholars among the top 100 authors. Technology investment and innovation achievements are essentially mismatched. The former has become much sought after, and project and research funds have become rent-seeking space for some authorities, generating academic corruption to some extent. The utilitarian and instrumentalized view of technology has emerged and spread in the technology community, and academic fraud has been universal. An innovation-oriented mainstream value cannot be formed and widely accepted.

4.2. Incentives for new subjects

In the market economy, enterprises are supposed to be the subjects of technological innovation. They should gradually shift from imitative innovation to indigenous innovation based on their conditions if they want to continuously enhance their market competitiveness in the process of technological innovation. According to statistics, however, enterprises with independent intellectual property rights and core technology only account for 0.3‰ in China,[7] and many researches have proved and will continue to prove that technological capabilities, especially core technological capabilities, cannot be bought and can only be developed through learning and indigenous innovation. Technology learning is organizational, and thus it cannot be initiated without the vision and determination of strategic decision makers. So, under what conditions is it most likely to produce such vision and determination? Organizational behavior research proposes that only with relatively high ambition can an organization have motivation to develop new abilities through learning. In the market competition, the crucial reason for the change in corporate ambition is crisis awareness, which assumes relevance when an enterprise fails to deliver expected performance. Do Chinese state-owned enterprises, private enterprises and corresponding research institutes have sufficient crisis awareness that will prompt them to strengthen organizational learning of technology and indigenous innovation for active reform?

---

[7] *Ibid.*

(1) **State-owned enterprises:** From the perspective of economic system reform, China has implemented planned economic system before reform and opening-up, and this system fundamentally excludes market mechanism. State-owned enterprises have neither the autonomy nor the motivation and pressure for technological progress. Their initiative of technological innovation has been constrained, thus severely influencing their technological progress and indigenous innovation. Despite this, China has intensified reform of state-owned enterprises in the recent years, the inertia brought about by the planned economic system adopted over the years has forced a high proportion of large and medium-sized state-owned enterprises allocate factors through administrative power. As a result, their impulse for technological innovation cannot be truly stimulated. From the angle of authorization chain and agency relation of state-owned enterprises, their management has the characteristics of two kinds of agencies: administrative agency and economic agency. For the former, enterprise is a subsidiary of the government; for the latter, enterprise is the independent subject of market competition. Due to the existence of such dual identities and vague roles, leaders of state-owned enterprises seek only short-term profits while abandoning long-term technology R&D and dissemination. What is even more serious is that those leaders specially emphasize the special relationship with government bureaucrats and spend more energy on non-technological innovation activities, such as lobbying superiors and building interpersonal networks.

(2) **Research institutes:** Research institutes represent the most dynamic and productive knowledge economy and are one of the major contributors to indigenous innovation. In order to increase national competitiveness and innovation strength, Chinese government started reforms centered on reducing appropriation for undertakings of research institutes in 1985 and initiated a transformation plan for research institutes in 1999, forcing research institutes to be market-oriented. A small number of research institutes have been successfully transformed and have achieved expected results. However, for most, the transformation violated basic laws of coordination between technology and economy, as it required commercialization of technological activities in advance, especially by taking mandatory measures within limited time. In this way, research institutes have undertaken reforms to realize short-term economic benefits, which undermined

the operating mechanism of the large circulation of technological economy. On the one hand, years of accumulated technological experience and research achievements have been diffused with the implementation of transformation, and many research achievements have gone to the MNCs through the flow of researchers and have been rapidly transformed into real-world technologies and products, further widening the gap between domestic enterprises and MNCs. On the other hand, research institutes tend to be more secluded, and the few with fruitful achievements are more inclined to work on papers behind the closed door. The macro-mechanism that combines technology with economy has been continuously undermined, and the indigenous innovation capacity has further declined.

(3) **Private enterprises:** Private enterprises are the product of the reform of China's economic system and have a short period of development in China. In the early stage of development, private enterprises were suppressed. With regard to scale and structure, private enterprises are far from being comparable to state-owned enterprises and foreign-funded enterprises, but they have unique advantages in technological innovation, as they have strong innovation impulse, flexible innovation methods, and rapid innovation decisions for survival and development. A large quantity of private enterprises have played an increasingly significant role in technological innovation. Yet, the source of indigenous innovation of private enterprises is mainly internal. While providing an important foundation for continuous emergence and survival, this is destined to affect its further development. The main reason is that there are obvious information barriers that prevent effective and large-scale collection of information, such as market conditions, new technological development, funding sources for technology development as well as government policies. Private enterprises lack adequate human resources and on the whole face the deficiency in following the trend of new technology and cultivating awareness of opportunities. Meanwhile, the shortage of funds is also viewed as an obstacle to indigenous innovation of private enterprises. Since indigenous innovation has higher risk and the process starting with technological achievements and ending in market is longer, adequate capital is necessary. In addition, any pioneering conception, design or implementation is full of uncertainty with unpredictable results. Yet, restrained by factors, such as relatively weak capital strength, shortage of technical talent, and poor access to technical and

market information, private enterprises face quite low probability of success and need to undertake higher risks. This is particularly serious in R&D of high and new technology. These obstacles determine that the indigenous innovation of private enterprises is more about innovation of applied technologies, including improvement of processes and products. It is difficult for them to make extensive breakthroughs in core technology.

4.3 *Policy support and constraints*

Be it developed countries or emerging market countries, almost all major technological innovations are strongly promoted and supported by the government. The governmental support for corporate indigenous innovation is evidenced by macroeconomic policies, such as creating a sound environment for social and economic operations and encouraging corporate indigenous innovation. It should mainly include the following three aspects. First, the government can improve production factor market and establish a production factor price system that accurately reflects the scarcity of production factors. Enterprises can combine new production factors through technological innovation to reduce production costs and improve profitability. Second, the government should increase investment in basic scientific research, establish technical innovation support system that integrate enterprises, universities and research institutes, which is conducive to forming an effective mechanism for smoothly transforming technological innovation achievements into corporate productivity and promoting indigenous innovation. Third, the government should maintain sound market competition order and enhance protection of intellectual property rights, enabling enterprises to enjoy economic benefits brought about by indigenous innovation and to cultivate internal incentive mechanism for indigenous innovation. Fourth, the government should implement fiscal and tax policies that encourage enterprises to become the subjects of technological innovation. A set of fiscal and tax systems that encourage indigenous innovation of enterprises should be set up to tackle the problem of difficulties in financing for innovation and encourage enterprises to develop self-owned brands.

China's unimproved factor market, imperfect public service institutions and poorly implemented protection for intellectual property rights constitute obstacles to technology transfer and diffusion of foreign-invested enterprises. Meanwhile, domestic product market structure impedes the diffusion of new technology within the industry and the generation of technological innovation. It is mainly manifested in the following aspects. First, since the domestic financial market is not open enough, there are fewer financing channels of enterprises, especially small and medium-sized enterprises, leading to a very high cost of technology introduction. Second, free flow of personnel is faced with many constraints, the flow and entrepreneurship of technical management personnel are immature, and channels of technology spillover are not smooth enough. Third, product market is imperfect, and monopolistic industries and administrative controls still exist in a certain scope. All these result in redundancy in technological level between MNCs and local enterprises, which influences the speed of technology transfer and diffusion within MNCs and is disadvantageous to the technological progress of local enterprises in China. Fourth, due to the imprint of China's political system and planned economy, the formulation of macroeconomic policies is also less in line with the laws of economic operation. The governments at all levels have more or less unsatisfactory service awareness, mainly manifested by the commander awareness in decision-making of indigenous innovation and administrative intervention in actual operation. Of course, it should not be rejected that the state and governments offer major investment and special support to key technologies and industries, such as two bombs and one satellite, Shenzhou spacecraft and other projects of national defense significance. The government should provide sustained and effective policy support and fiscal technology investment. Indigenous innovation of most technologies, however, should and must be driven by market mechanism. Indigenous innovation in the automobile industry in China has met obvious administrative intervention from the government. The automobile industry policies implemented over the past two decades have not only failed to enable China's automobile industry to acquire the ability of independent development but also deprived it of the original development capability. As a result, the industry missed the historic opportunity of development and left endless troubles that lead to immeasurable losses. In the recent 5 years, however, the Chinese government has gradually opened the market in accordance with the requirements of the WTO, allowed competition between domestic enterprises and MNCs, and offered necessary support. Thus, the indigenous innovation capacity of local enterprises has improved rapidly.

Chapter 9

Policy Suggestions for the Enhancement of Indigenous Innovation Capacities of China

1. Macro-policy Suggestions for Governments in China

1.1. Encourage and attract MNCs' investment in R&D internationalization

By analyzing the R&D internationalization of MNCs, we find that it brings extensive spillover effects on the technological progress and indigenous innovation of host countries so that the technological efficiency of local enterprises will increase as cross-national competition intensifies. Although it is impossible for China to attract MNCs' investment in R&D internationalization for all technological fields, it is possible to attract such investment on a large scale in some fields. Therefore, in actual operation, it is necessary to achieve the goal of encouraging and attracting MNCs' investment in R&D internationalization through means such as strategic design and policy arrangements. In terms of location distribution, China should adhere to the principle of "classified guidance and gradient progress", follow the objective law of non-balanced development, proceed from reality, adapt to local conditions, and continue to give full play to regional advantages so as to attract more MNCs to set up R&D institutions in China, localize their technologies. and bring advanced operational modes and management methods into China. Specifically, this can be achieved mainly in the following aspects.

1.1.1. *Increase policy support*

Various policies and measures should be adopted to encourage and attract MNCs' investment in R&D internationalization. In addition to refining the laws and regulations related to MNCs' investment in R&D internationalization, China should legally safeguard the normal operation of foreign-funded enterprises and ensure that their legitimate rights and interests are not infringed. While providing a good investment environment for the R&D of MNCs, China should focus on three aspects, namely, taxes (tax rate for all aspects of MNCs' R&D investment and management should be appropriately lowered), financing (guarantee conditions for financing in China by MNCs for R&D investment should be less strict) and insurance (domestic insurance organizations may provide R&D institutions invested by projects that are encouraged by China with insurance on political risk, performance and guarantee), to support the R&D activities. In addition, China should improve the building of service system for R&D of MNCs, actively help MNCs to solve practical problems they encounter during R&D, and improve its attractiveness to MNCs in terms of soft environment for R&D investment.

The approval system for foreign-invested projects involves setting up Sino-foreign R&D institutions. During the introduction of Sino-foreign R&D projects, China must adhere to the principle of mutual benefit and the projects must be beneficial to China. Moreover, such consideration must be given from the corporate as well as the industrial perspectives. Major projects of key departments of the industry, especially those monopolized throughout the industry, must be controlled by China. R&D projects with advanced technology, vast domestic market and good economic returns are not suitable for joint ventures. Even if joint ventures are required in specific cases, China must have the control.

1.1.2. *Adhere to regional "non-balanced" development*

Based on the measurement of the level of R&D internationalization in Chapter 2, it can be found that the level is also non-balanced due to the significant differences in China's regional economic development. For any developing country, it is impossible to attract direct R&D investment from MNCs for all technical fields. China is no exception. China is a large developing country. There are huge differences in the types and fields of technology as well as in the overall technological level across regions. Therefore, it is necessary to build a mechanism that allows the rational use of R&D resources between regions and promotes domestic technological progress. China should follow the development trend of MNCs' investment in R&D internationalization in China. In terms of location distribution, China should follow the principle of non-balanced development and attract MNCs to establish R&D institutions extensively in coastal provinces and cities, and appropriately in the central and western regions. Western central cities, such as Xi'an, Chongqing, and Chengdu, should encourage and attract MNCs to set up R&D institutions in areas where they have comparative advantages, such as deep processing of resources, automobiles and chemicals. The east-to-west R&D chain, which consists of Beijing, Shanghai and Shenzhen, is just like the global "highland for innovation", such as Silicon Valley, Bangalore and Hsinchu. It has become China's core innovation zone. Therefore, it is essential to research into the key fields of these three cities. A prominent feature of China's economic growth is the imbalance of regional economic development. Each region has distinct characteristics in their modes of economic growth. Therefore, a rational system for division of labor among the regions should be built, in terms of distribution of industries and countries, based on regional characteristics in resource endowments, key industries that attract R&D investment, and space for expansion.

1.1.3. *Refresh understanding of the significance of attracting foreign investment*

From "market for technology" to encouragement of technology spillovers, the positioning of attracting foreign investment has undergone some changes over more than 30 years of reform and opening-up. The distribution of MNCs' R&D institutions in China is based on the layout of the production system. Therefore, for local governments, formulating measures to attract more MNCs to invest in the region is the key to attracting them to make local R&D investment. In general, the more productive investments an MNC makes in a region, the larger its scale of production is. Then the necessity and possibility of establishing an R&D center to support local production are greater. The higher technological level that the products of its subsidiary have, the more innovative its R&D activities will be.

When a country receives a large number of foreign investments, FDI will generally become a major factor for the host country to attract cross-border R&D investment. As MNCs' scale of investment in the host countries increases, their subsidiaries in the host countries will continue to expand their demand for technological R&D in order to adapt to the expansion of demand in the markets of the host countries. Therefore, the level of R&D internationalization will increase with the growth of foreign investment. China's focus should shift from introducing foreign capital to fill in the financial gap to attracting investment in R&D internationalization by introducing MNCs to make direct investment, thus obtaining the technology spillover effect of foreign capital.

To be specific, the significance of China's current attraction of FDI should be positioned to promote domestic technological progress, so as to better attract MNCs' investment in R&D internationalization. While attracting FDI, a country with a larger market size will more likely become the factor that attracts cross-border R&D investment. Market size is related to the speed and level of economic development. Currently, the world economy continues to be sluggish and there are still great uncertainties in the international economic environment. However, the Chinese economy is "outstanding" and has been growing rapidly. China's good macroeconomic environment decides the effective demand of market, giving rise to the constant expansion of market capacity. Therefore, when foreign demand is insufficient, further increasing fields for opening up and competition as well as continuing to stimulate foreign investors to invest in China undoubtedly have great appeal to MNCs in making R&D investment.

1.2. Focus on building a highland for innovative talent

1.2.1. Strengthen development of human resources for R&D

As revealed by the analysis of the motivations and selection of location for the R&D internationalization of MNCs in Chapter 1, MNCs attach great importance to the availability of professional technicians in the host countries when selecting overseas locations for R&D. Therefore, high-quality human capital is a major factor that attracts MNCs to conduct R&D activities. The fundamental element of technological progress is talent, which is why Schultz proposes the concept of human capital. The training, attraction and use of talent form the key to the development of science and technology in a country as well as provide an important location advantage for attracting MNCs to make high-tech investment and conduct R&D activities. In addition, only when the local human capital has higher quality and is capable of solving some substantial problems will there be opportunities to access advanced technologies during cooperation. In this way, the effect of "learning by doing" will increase, thus helping developing countries to acquire substantial technologies during cooperation and improve the ability of local enterprises to digest and absorb technologies.

China still lags behind in the development of human resources and faces serious brain drain. Therefore, it should pay attention to the development of human resources. In particular, it should deepen the reform of the education system and the training system for corporate personnel and train professionals with strong capabilities and the pioneering and innovative spirit. It should also actively introduce foreign professionals who master the world's cutting-edge science and technology while retaining the overseas returnees and giving full play to their talent with effective policies and measures.

1.2.2. *Refine the mechanism for training innovative talent*

First, the current education system should be reformed to ensure the overall quality of talent training. Spoon-feeding teaching should be replaced by heuristic and interactive teaching, which should start with education for the children. It is necessary to cultivate children's innovative spirit, enhance their hands-on ability, and encourage them when they make small inventions. In contrast, currently, higher education also has many shortcomings. The major ones are as follows. The teaching of professional knowledge fails to keep abreast of the times or the frontiers of industrial technologies. As a result, there is a gap between the professional knowledge students have learned and real-life technologies. The blind enrollment expansion of colleges and universities, especially for graduate education, has led to serious shortage of teachers. In addition, due to the low thresholds in enrollment and graduation, the ability of students is greatly reduced. Graduate students, especially those majoring in science and engineering, have a lower professional level. Colleges and universities should guide students to make cutting-edge exploration and ensure the overall quality of Chinese researchers while avoiding blind enrollment expansion.

Second, social resources should be rationally developed to foster awareness of innovation. The whole society should create an atmosphere that encourages innovation and an innovation mechanism. An innovation fund should be established. Encouraging the awareness of innovation is first of all reflected in the tolerance of mistakes. In particular, during development of some applied technologies, there should not be rush; more space should be given to the project researchers and developers so that the purpose of training innovative talent can be truly achieved on the basis of trial and error.

Third, the current monotonous talent training system should be improved. At present, innovative professionals in China generally depends only on the professional training in schools. Therefore, it is often difficult for them to adapt to the needs of the actual R&D behaviors of enterprises. Therefore, it is necessary to strengthen the mechanism for cooperative training of talent between domestic universities and MNCs. For example, Microsoft Research in China selects 10 outstanding Ph.Ds. in computer science from universities each year, awards them the title "Microsoft Scholar", and funds their overseas academic activities. Its "Microsoft Scholarship Program" has greatly promoted the training of outstanding local software professionals in China.

1.2.3. *Train innovative talent under R&D strategic alliances*

Among the behaviors of MNCs' R&D internationalization, R&D strategic alliances are a special form, with certain peculiarities in talent training. On the one hand, training of talent for R&D strategic alliances should attach more importance to internationalization. Specifically, it is necessary to greatly improve the local talent composition for the R&D alliances of MNCs in China and fully attract foreign high-tech professionals to work in the R&D bases in China. Information-based means should be fully utilized for global sharing of talent resources and greater accessibility of international talent resources. On the other hand, incentive strategies for innovative talent should be highlighted. Among various reward and incentive strategies, option incentives are most suitable for professionals of multinational R&D strategic alliances. For option incentives, alliances and developers are not only viewed as a kind of entrustment-agent relationship, but a kind of equity (share) trading relationship with assets as the link. This changes the relationship of managing and being managed. Instead, a kind of community of shared interests is formed, with alliances and employees (technology professionals) forming a strategic partnership. Option incentive, on the one hand, mitigates the risks for the shareholders and employees of the alliances and fixes the conditions for transaction, and on the other hand, focuses on the long-term interests of the alliances and allows the realization of individual interests through the development of the team, which is conducive to unlocking the employees' passion for work. Besides, option incentive focuses on the portion of new growth, thus with no harm to the interests of the alliances or the established owners of the enterprises. It is a win–win strategy.

1.3. Help local enterprises acquire the status as subjects of R&D innovation activities

Enterprises are the subjects of economic operation and of digestion and absorption of technology spillovers brought about by the R&D internationalization of MNCs. To cope with the challenges of R&D internationalization of MNCs, China must fundamentally reform the national innovation system, gradually change the pattern with the country as a single investor of scientific research, and establish a technological innovation system with enterprises as the subjects of innovation. Despite years of reform of the enterprise system, a large number of state-owned enterprises in China have not yet become modern in the true sense, as manifested by their inadequate pursuit of technological innovation, which leads to insufficient direct motivation for technological development and for R&D investment. The situation that China's social R&D funds are over-reliant on government grants has not been fundamentally changed. The pursuit of stronger economic competitiveness through technological progress has not become a conscious act. Therefore, China should actively implement the *Decision of the Central Committee of the Communist Party of China on the Reform and Development of State-owned Enterprises*, establish a modern enterprise system with clear property rights, clear powers and responsibilities, separation between government and enterprises, and scientific management, and form a technological innovation system centered on enterprises. It is necessary to help enterprises acquire the status as subjects of "self-reform and self-development" in technological development under the market economic system, so that enterprises can continuously intensify technological transformation and development, as well as create a good micro-environment for introducing, digesting and reforming technologies. China should accelerate the transformation of the business operation mechanism, secure enterprise status as the subjects of technological innovation, make enterprises the subjects of investment, risks and interests in technological innovation, and realize the transformation of state-owned enterprises from traditional production-based enterprises to innovation-based ones. Enterprises are the basis for technological progress in developing countries. The creation of dynamic micro-subjects adapted to the market economic system and the modern enterprise system is a key factor influencing technological progress and is an urgent and arduous task.

1.4. Optimize the environment for R&D internationalization of MNCs

1.4.1. Establish and improve the supporting service system for R&D

The service environment should be improved in an all-round way, with a sound service system supporting R&D internationalization activities of MNCs. For instance, R&D personnel of MNCs may be provided with more convenient conditions, industrial associations that serve as the link, and accounting firms and highly professional law firms in place of audit agencies. The infrastructure environment should be improved so that the R&D institutions of MNCs can focus more on their operations and innovation activities, thereby contributing to the effectiveness of technology spillovers.

Due to a late start of China's service industry, relevant policies and regulations lag behind those of the developed countries and fail to meet the needs of R&D institutions of MNCs. Therefore, the government should assume an active guiding role, improving market regulations on the one hand and boldly decentralizing powers on the other hand to simplify the exit–entry approval procedures for R&D personnel of MNCs. The "green card" system for talent should be promoted and the problem of household registration for employees of MNCs should be addressed. In this way, a good environment is provided to the development of the service industry, with a series of service organizations improved to pave the way for the development of foreign R&D institutions. The same is true for the construction of the infrastructure environment.

1.4.2. *Strengthen the construction of basic communication and information facilities*

Judged by attributes, R&D activities involve high risks and high input, and often require constant communication between the internal production departments and R&D institutions that are located in different places. Thus, they tend to have higher requirements on the quality of the regional infrastructure. This is especially important in attracting MNCs' R&D institutions. Good infrastructure, especially in terms of regional communication and information, is a prerequisite for better development and utilization of information resources. The impact of information infrastructure on R&D institutions is mainly reflected in the R&D efficiency of MNCs. As the product development cycle continues to shrink, the competition between MNCs, especially the world's top players, is in a sense about the winning of opportunities for technological development and first-mover advantage for effective diffusion on a global scale. Therefore, for them, the time delay in information transmission is fatal. When selecting locations, MNCs pay more attention to information infrastructure to ensure timely and efficient information flow and personnel exchange. The communication and information infrastructure provides a large number of information and communication platforms for MNCs' R&D institutions, which allows them to reduce their costs on acquisition of information and transportation, maintain close contact with the headquarters, and capture international and domestic market information and achieve external economies of scale, such as agglomeration economies. These are conducive to the guidance, transformation and dissemination of technological innovation. Therefore, a location with good infrastructure will have a positive impact on the entry of the R&D institutions of the MNCs.

At present, the R&D institutions established by the MNCs in China are mostly found in Beijing and Shanghai because they offer better information infrastructure. However, there is even a certain gap between Chinese cities which have the best information infrastructure at home, namely, Beijing and Shanghai, and the advanced countries and regions. As MNCs make deployments across China, setting up a certain number of R&D institutions in the central and western regions in China is an inevitable choice. The infrastructure conditions in these regions are a major obstacle to the MNCs' choice for R&D investment. Now, their requirements on the infrastructure environment are no longer simply about whether there are remote-controlled telephones and comfortable office conditions. Currently, R&D activities are global and usually involve round-the-clock work. A large amount of data need to be transmitted and an efficient information network has become a must. Therefore, strengthening the construction of information infrastructure is a major guarantee for attracting R&D internationalization of MNCs.

1.4.3. *Guide the construction of technological infrastructure for domestic R&D institutions*

The R&D internationalization activities of MNCs do not subjectively generate spillover effects. Hence, how to better absorb the objective spillover effects of R&D internationalization of MNCs is a very important problem, for which the construction of technological infrastructure, like domestic R&D institutions, will be a fundamental and important task. In the knowledge economy, technological infrastructure is a prerequisite for R&D, efficient production, and full use of technological achievements. Such infrastructure includes the general or basic technologies that form the core technologies of an industry and serves as the basis for the development of products, processes and services in the industry. Another type of infrastructure is called technological tools, with which enterprises make some economic behaviors more efficient or more likely. Study on the Organization for Economic Co-operation and Development (OECD) shows that, for R&D internationalization behaviors, MNCs tend to choose locations with advanced and complete R&D infrastructure, so that they can obtain more comparative advantages than competitors in technological innovation and improvement of productivity. The type of public investment is increasingly becoming an important part of a country's competition strategy because it can not only affect the growth of the new technology economy but also attract more private investment than non-current economic assets. Thus, it has a positive influence on the domestic economy. China's increase in investment in technological infrastructure and improvement of R&D infrastructure, including establishment of a fair number of R&D institutions, are conducive to attracting MNCs to invest in R&D. The government can also develop specialized high-tech parks for R&D in a particular industry to promote the development of R&D institutions associated with the industry.

1.5. *Improve the system for protecting intellectual property rights*

1.5.1. *Fully understand the significance of protecting intellectual property rights for R&D activities*

R&D activities of MNCs are the process of knowledge production, with intangible products. In addition, technology spillover is an objective phenomenon of R&D activities. Therefore, when MNCs are engaged in R&D internationalization activities, they are very concerned about the protection of intellectual property rights of R&D achievements. The system of intellectual property rights is an institutional arrangement that safeguards the subjects of innovation when they obtain the benefits of innovation and plays a very important role in promoting innovation, technological progress and economic growth. Besides, a sound system for the protection of intellectual property rights is a sign of the maturity of market economy. It is also an important determinant of the choice of location for R&D internationalization by MNCs. In the recent years, with the advancement of market economic system, China has begun to attach more importance to the protection of intellectual property rights and has also formulated relevant laws and regulations. It can be said that China has made substantial progress in legislation for the protection of intellectual property rights. The *Patent Law* and the *Trademark Law* have been revised several times. The protection of intellectual property rights such as patents, trademarks and trade secrets has met the TRIPS regulations of WTO. In addition, China has also participated in all important conventions on intellectual property rights. However, China's laws and regulations on protection of intellectual property rights in China are still not perfect compared with those of developed market economy countries. Those already enacted, in particular, still face big problems in enforcement, which affects the investment of enterprises in R&D activities. Therefore, the Chinese government should further strengthen the promotion of protection for intellectual property rights, especially improving the awareness of domestic enterprises of protecting intellectual property rights based on the changes in the landscape of international competition after China's accession to the WTO. China should increase penalties on infringements of intellectual property rights and improve and refine the protection system for intellectual property rights. The rights and interests of foreign-funded enterprises concerning R&D achievements should be protected from infringement in accordance with law, including technical secret, transfer of property rights, and prevention of counterfeiting and infringement. However, for net importers of intellectual property rights, excessive governmental protection is tantamount to falling into a pit of one's own digging, which hinders technology diffusion inside the countries. Therefore, it is important to find an equilibrium between effective protection and full competition, and balance benefits between technology creators and users.

1.5.2. *Improve the system for managing intellectual property rights*

The basic system for managing intellectual property rights in China works as follows. China National Intellectual Property Administration administers patents, the National Copyright Administration under the General Administration of Press and Publication administers copyrights, and the Trademark Office of the State Administration for Industry and Commerce administers trademarks. The courts and the departments of public security, customs, and industry and commerce are responsible for law enforcement. In addition, the regulatory authorities of some products are the drug supervision and quality inspection departments. The decentralization of power in various departments has produced many negative effects, especially on the protection of intellectual property rights in the legislative and judicial stages. Therefore, it is necessary to improve the management system for intellectual property rights. Specifically, the national departments of intellectual property rights should be coordinated and adjusted to integrate patent, trademark and copyright for internal and external standardization. After the integration, national management departments of intellectual property rights should be the administrative department at the ministerial level directly under the central government. The provincial management departments of intellectual property rights should also be standardized.

1.5.3. *Strengthen the promotion and law enforcement of intellectual property rights*

A good cultural atmosphere for protecting intellectual property rights should be fostered, which requires long-term persistence of instilling the concept of such protection into relevant enterprises and individuals. Therefore, the government may consider educating party and government leaders, enterprises, institutions and individuals by organizing training courses and issuing training materials, especially those on the information and knowledge about international protection of intellectual property rights. At the same time, attention should be paid to the enforcement of relevant protection laws. At present, the most important problem of China's protection system for intellectual property rights is not the absence of laws but the refusal to obey them. In the past, punishment on violations of intellectual property rights was not strict enough without deterrence on the infringers. From now on, it is necessary to increase the punishment. In addition to fines, it is also necessary to consider suspending business and canceling licenses. Finally, the domestic laws should be aligned with international ones and coordinated with the agreements on intellectual property rights in the WTO rules. Therefore, China should continue to improve and strengthen the enforcement of laws on intellectual property rights so that there are laws that people can and must obey.

1.6. *Improve the R&D factor market and product market*

R&D Internationalization of MNCs, as a production activity of knowledge products, requires input of factors and output of knowledge products. Therefore, for this production activity, the host countries need to make purposeful adjustments based on the input factor market and the output product market. In other words, the government should improve the factor market, especially the capital market and the labor market. For R&D activities, the main input factors are R&D capital and R&D technicians. In terms of R&D capital, it is necessary to further develop and open up the domestic financial market and expand financing channels for enterprises and individuals, such as developing personal loan guarantee and venture capital funds, thereby reducing the cost of technology introduction. In terms of R&D technicians, it is necessary to reduce the regulation of personnel flow and encourage the flow of technology managers and entrepreneurship, clearing the channels of technology spillovers. A country that has higher labor quality will attract more technology-intensive and capital-intensive MNCs. For developing countries, increasing investment in human capital, on the one hand, can increase labor productivity and directly promote economic growth. On the other hand, it can promote physical investment, attract capital-intensive MNCs, and improve the effectiveness of local enterprises in absorbing advanced technologies. This requires developing countries to increase investment in education and strive to improve their human capital level. The government should strengthen the formal education in schools and the non-formal education in the society, reform the education system, adopt the path of cooperation between enterprises, universities and research institutes, train high-tech talent needed for national development, and enhance the country's ability to absorb and innovate technologies. Besides, it should improve the product market to foster a fierce competitive environment that drives MNCs and local enterprises to improve their technological level and accelerate the technology transfer and diffusion within MNCs.

2. Micro-reaction Suggestions for Enterprises in China

2.1. *Improve ability to innovate*

2.1.1. *Expand R&D investment and improve international competitiveness*

In the context of economic globalization, the fundamental element of international competition among enterprises is the technological innovation capacity, whose improvement can first of all be an effective means to dealing with the technological competition from MNCs. It also helps local enterprises to better absorb the technology spillover of R&D internationalization of MNCs. In the past, the proportion of R&D investment by Chinese enterprises was very low. Basic technology research was basically undertaken by state-owned research institutes. Enterprises are generally only engaged in R&D work concerning industrial development and process technology. Faced with the increasing pressure from MNCs' R&D investment in China, Chinese enterprises should increase the intensity of R&D investment and improve the industrial transformation capacity of technologies. At this stage, the R&D investment of Chinese enterprises cannot be excessively increased. It must be made within the capacity of enterprises. The state should be responsible for basic research and give preferential policies and support to the indigenous innovation of technologies of enterprises. Major enterprises in key industries should be encouraged to make cutting-edge technological innovations so as to break through the "technology lock-in" of MNCs. In view of R&D internationalization of MNCs, in addition to increasing R&D investment, enterprises should choose appropriate strategic models for technological innovation based on their own strength, the landscape of international competition and the overall business objectives. Leading enterprises of domestic industries should make full use of the historical opportunities brought about by the R&D internationalization of MNCs, keep up with the development trend of international technologies, increase R&D investment, improve their technological innovation capacity, and strive to become international technology leaders. Enterprises that are not strong in technology should actively cooperate with MNCs, start with imitative innovation, introduce the MNCs' technologies, and transform them into their own capabilities through learning, digestion and absorption, thus continuously improving their level of technological innovation.

2.1.2. *Improve enterprises' capabilities of digestion and absorption*

During the advancement of technological innovation capacity of enterprises, digestion and absorption form the starting point. They are the strategic behaviors of enterprises for internalizing external knowledge and materializing it into products. In general, Chinese enterprises' capabilities of digestion and absorption are not strong but are improving. As mentioned earlier, Chinese enterprises have increased their R&D investment, which improves their indigenous innovation capacity as well as capability of absorption. Cohen and Levinthal (1989) propose the dual roles of R&D activities through theoretical and empirical analysis. The first role is innovation, i.e., the generation of knowledge through the enterprise's own efforts. The other role is learning, i.e., obtaining knowledge of competitors or other industries. This is the concept of the capability of absorption. They further point out that such capability includes three aspects: understanding new knowledge in the external environment, digesting and absorbing external knowledge, and applying it to business to achieve commercial purposes.

During the R&D internationalization of MNCs, as the complexity of knowledge increases, the capability of absorption becomes very important. The failure of China's market-for-technology strategy is largely due to the fact that Chinese enterprises only know about introducing technologies and ignore the development of capabilities of digestion and absorption. According to the theory of technological gap, Chinese enterprises can narrow the gap between them and MNCs by increasing R&D investment, thus speeding up the technology transfer of MNCs. Based on model analysis, Wang *et al.* (2004) find that there is only one Nash equilibrium to the technological gap between MNCs and manufacturers of the host countries. If the technological gap is greater than the equilibrium, technology transfer of MNCs will slow down; otherwise, it will accelerate. The technological gap between MNCs and manufacturers of the host countries determines the speed at which MNCs transfer technologies to their subsidiaries.

2.1.3. *Expand channels of corporate technological innovation*

Chinese enterprises have the following tendencies in technology acquisition. First, they simply understand technology acquisition as the introduction of technology that does not require their own technological efforts. Second, among the technologies introduced, complete sets of equipment and key equipment dominate. Third, more closed independent development is their internal effort on technology acquisition. These tendencies are manifested in the superficial perception of technology acquisition and the monotony and rigidity of acquisition methods. To address them, Chinese enterprises should flexibly use two major approaches: market-based acquisition and non-market-based acquisition. At a time when the Internet is widely used, the access to free-flowing information may yield twice the result with half the effort. Technical information searchers are indispensable in technology management within enterprises. More importantly, the access to "wetware" based on acquisition of technical talent has more positive externalities in technology acquisition. Enterprises should not be satisfied with the relatively closed independent R&D but should focus on methods of external technology acquisition like dynamic establishment of cooperation agreements.

2.2. *Build a development model of integration of enterprises, universities and research institutes*

2.2.1. *Strengthen cooperation among enterprises, universities and research institutes*

With the deepening of reform and opening-up, state-owned enterprises have continued their reform and the private ones have enhanced their strength. But their R&D capabilities are still weak. In contrast, domestic universities and research institutes have stronger research capabilities.

Therefore, enterprises can strengthen cooperation with them and improve technological R&D by virtue of their scientific and technological strength. For domestic universities and research institutes, they can also solve problems, such as venues and funds. The two sides can make up for each other's deficiencies and learn from each other.

However, there are still many problems in real life. Although the development model of integration of enterprises, universities and research institutes is beneficial to both supply and demand sides of technology, not many enterprises have achieved this, considering the feasibility of projects and the applicability of scientific and technological achievements. Product R&D takes a certain amount of time, but enterprises often do not have patience with scientific and technological achievements that cannot bring immediate benefits. They have no intention of long-term cooperation with research institutes, thus missing many new achievements.

In addition, be it MNCs or local enterprises, their strategic alliances for joint R&D established in China are mostly limited to cooperation with domestic universities in the field of basic research. The main reason is that most Chinese enterprises have weak R&D capabilities or even have no technological innovation capacity, which makes it difficult for them to be aligned with international R&D activities. Domestic universities and research institutes have advantages in scientific research capabilities, so they maintain a wider range of technical cooperation with foreign MNCs. Under this circumstance, enterprises may use universities and research institutes as a window to the development trend of international science and technology, and indirectly establish communication with the R&D activities of the MNCs through cooperation and exchanges with domestic universities in R&D. In terms of specific measures, China may learn from some practices of the United States, such as establishing research centers for cooperation between universities and industries. These research centers can take many forms and receive funding from the government. In addition, personnel exchange programs between universities and industries should be encouraged to have university researchers conduct research in enterprises.

2.2.2. *Explore the integration of enterprises, universities and research institutes*

The development model of integration of enterprises, universities and research institutes can not only be applied to domestic cooperation but also go beyond borders for multinational cooperation among enterprises, universities and research institutes, which is of great significance to Chinese enterprises, whose technologies are lagging behind. The European Union's European Strategic Programme on Research in Information Technology (ESPRIT) is a typical example of such integration. In the early 1980s, the program was developed by the European Commission, which was then responsible for industry, to bring together large and small electronics and information enterprises in the EU to work with universities and other research institutes to tackle problems and improve the EU's technological capabilities in the field of electronic information. The program was divided into four phases from 1984 to 1998. A total of 1,000 R&D projects and 900 auxiliary projects were implemented, yielding several thousand achievements. Such significant achievements have made major contributions to the improvement of competitiveness of the EU's information and communication industry. Promoting cooperation among enterprises, universities and research institutes is the solution to optimizing resource allocation, facilitating the transformation of scientific and technological achievements into productivity, and improving the technological innovation capacity of Chinese enterprises. The cooperation among enterprises, universities and research institutes is essentially a network of knowledge creation and diffusion. Since knowledge is sticky, how to establish an efficient cooperation mechanism, such as benefit distribution and risk-sharing mechanisms, is a crucial problem confronting the cooperation.

2.3. Implement the strategy of going international for local enterprises' R&D

Since reform and opening-up, a major problem that constrains China's economic development is that the technological level and infrastructure for technological innovation lag behind. Improvement of technological level allows the rational use of international R&D capital in addition to making full use of domestic R&D capital. In other words, the strategy of going international should go beyond sale and production to the R&D process. Therefore, the powerful and well-equipped Chinese enterprises can choose to make R&D investment in countries with a good environment for technological innovation, make use of global R&D resources and establish their own global network for technological innovation.

This allows them to keep track of the development of the world's cutting-edge technologies, accumulate technologies, improve efficiency in technological innovation and shorten the development cycle of technology. It can also break the international technology lock-in and acquire advanced technologies that are difficult to introduce through other channels. To be specific, the methods detailed in the following sections may be used.

2.3.1. *Participate in the study and formulation of international industrial and product standards*

By participating in the activities organized by the international standards organizations, enterprises can learn about and master technological developments in the field in time as well as understand the needs of the market and users, thus avoiding detours in overseas R&D activities and shortening R&D time. On the other hand, they can also export their advantages in the process of product development to the international standards organizations and secure their position and technological advantages in the field so as to gain discourse power in product positioning in the international market. For example, in April 2001, ZTE was granted the independent membership of the international standardization organization 3 GPP2 to engage in the research into the third-generation mobile communication standard CDMA2000. This marks that ZTE has a say in the technological R&D for global third-generation mobile communication standards.

2.3.2. *Go international and monitor technological developments of technologically leading countries*

Since most Chinese enterprises do not take up a leading position in the world in terms of technological level, it is possible for them to test technologies in countries with more advanced technologies through the strategy of "going international" and observe the technological trends of competitors. The direction of technological development of relevant fields should be monitored to provide information guidance for R&D localization. This is an important means of making up for the insufficiency in capabilities during the transformation from imitation to innovation and is the precursor of innovation globalization. At present, Haier Group has set up information centers in 10 countries and regions, including Los Angeles in the United States, Lyon in France, Sydney in Australia, Tokyo in Japan, and Taiwan in China. These centers offer it strong support as they allow it to acquire timely international information about science and technology, market, design trends and laws.

2.3.3. Try to set up overseas R&D institutions

Similar to the level of research, the R&D environment and factors in China are also lagging behind. Therefore, enterprises can make indigenous innovation through the rational use of foreign R&D links and factors. Although Chinese enterprises are not a match for Western enterprises in both production scale and technological level, different from FDI, overseas R&D investment does not require enterprises to have monopolistic advantages but allows them to find new ways to make innovation. By setting up overseas R&D institutions, enterprises can respond to the special needs of different customers around the world and local production conditions as well as realize the localization of technologies. They can track and acquire technologies from host countries and competitors to benefit from the technology spillovers of local R&D, which can be a supplementary source of expertise. They can also get close to the knowledge centers, make use of the technological talent and research environment in the host countries to reduce R&D costs. In addition, for the achievements of development by overseas R&D institutions, patents can be directly applied for in the host countries to take the lead in occupying foreign markets.

2.3.4. Realize "localization" of R&D activities

The localization of R&D activities is a problem that must be highlighted when enterprises are going international in terms of R&D activities. The technological factors are the most complicated among the factors of production. They depend not only on the scientific organization and management but also on human factors, i.e., the researchers. Localization means localization of personnel, management and capital operation. When Chinese enterprises set up or acquire and merge R&D institutions overseas, it is particularly important to dispel the doubts of foreign R&D personnel on their managerial skills, corporate culture and local financing capabilities. They should adopt the same salaries, institutions, management structures, options and incentive mechanism as local R&D institutions to retain and attract outstanding overseas research forces. Compared with many MNCs, Chinese enterprises have a very low level of localization. After the international mergers and acquisitions, they still adopt the management model with control in the hands of the Chinese. It is very unreliable from the perspective of risk management that they have the Chinese who have little international experience to manage enterprises abroad. Therefore, only by analyzing the market situation in the host countries, rationally arranging R&D activities, and realizing the "localization" of R&D activities can the strategy of "going international" for R&D activities have sustainable and healthy development.

Of course, no matter what approach is chosen to "going international", the base for investment should be selected based on the purpose of the enterprises and the characteristics of the host countries. After their market share has reached a certain level, enterprises that have invested in overseas production should consider establishing R&D institutions in the host countries where consumption habits are more different from those in China, and implement the strategy of "going international" for R&D activities. Products suitable for the local market should be designed and developed based on the characteristics of local demands to further expand the market and technological space. Other enterprises with sufficient economic strength, especially the technology-intensive ones, should consider setting up R&D institutions in countries of competitors that are technologically advanced in the same industry to benefit from the technology spillovers of competitors.

2.4. *Establish international strategic alliances for R&D*

The strategic technology alliances are innovation-based organizations aimed at helping enterprises complement each other's technological resources and allowing them to together develop certain new technologies by reducing the development costs and risks of individual enterprises. The establishment of strategic alliances for R&D is conducive to the acquisition of foreign advanced technologies and skills by Chinese enterprises, the diversification of international business risks, the development of international markets, and the full play of comparative advantages. Cooperation and exchanges with the R&D institutions of the MNCs constitute an important way for Chinese enterprises to improve their technological innovation capacity. Through strategic alliances for R&D, Chinese enterprises often receive richer technical information for the enhancement of their research strength. Sometimes, MNCs provide strategic partners with technical data or prior technologies for the needs of overall development of technological innovation. This helps improve technologies. In addition, by establishing strategic alliances for R&D with the R&D institutions of the MNCs, Chinese enterprises can significantly improve their organizational and managerial capabilities and level for R&D. The potential impact of progresses in soft technologies will benefit the technological innovation of Chinese enterprises in the long run.

However, multinational strategic alliances for R&D are a form of organization that is extremely difficult to manage. Since the parties of alliances face problems concerning symmetry of benefit, balance of competition position, compatibility of strategic objectives and coordination of corporate culture, conflicts easily arise among them, leading to the failure of such alliances. For Chinese companies, to achieve success in alliance, strategic partners can be selected according to the 3C principle, namely, compatibility, capability and commitment. The 3C principle is a key condition for enterprises to find partners. If an enterprise's potential partners meet the 3C conditions, then the probability of success in working with them is greater. The ultimate goal is the development of corporate core capabilities, risk prevention, and internal coordination, trust, and communication.

Bibliography

Aigner, J., Lovell, K. & Schmidt, P. (1997). Formulation and estimation of stochastic frontier production function models. *Journal of Econometrics*, 6(1): 21–37.

Aitken, B. J. & Harrison, A. E. (1999). Do domestic firms benefit from direct foreign investment? Evidence from Venezuela. *American Economic Review*, 89(3): 605–618.

Aitken, B. J., Hanson, G. H. & Harrison, A. (1997). Spillovers, foreign investment and export behavior. *Journal of International Economics*, (43): 103–132.

Amann, E. & Virmani, S. (2015). Foreign direct investment and reverse technology spillovers: The effect on total factor productivity. *OECD Journal: Economic Studies*, 2014(1): 129–153.

Anselin, L. & Kelejian, H. H. (1997). Testing for spatial error autocorrelation in the presence of endogenous regressors. *International Regional Science Review*, 20(1–2): 153–182.

Archibugi, D. & Lammarino, S. (1999). The policy implications of the globalization of innovation. *Research Policy*, (28): 317–336.

Audretsch, D. B. & Feldman, M. P. (1996). R&D spillovers and the geography of innovation and production. *American Economic Review*, (86): 630–640.

Bai, J. H. *et al.* (2009). Application of stochastic Frontier model to evaluate the efficiency of China's regional R&D innovation. *Management World*, 10: 51–61.

Baldwin, R. E. & Forslid, R. *et al.* (2000). Investment creation and investment diversion. *NBER Working Papers*, 98(3): 423–438.

Banker, R. D., Charnes, A. & Cooper, W. W. (1984). Some models for estimating technical and scale inefficiencies in data envelopment analysis. *Management Science*, 30(9): 1078–1092.

Barros, P. P. & Cabral, L. (1994). Merger policies in open economics. *European Economic Review*, 38: 1041–1055.

Berger, N. & Humphrey, B. (1997). Efficiency of financial institutions, international survey and directions for future research. *European Journal of Operational Research*, 98(2): 175–212.

Bernstern, J. & Yan, X. Y. (1996). Canadian-Japanese R&D spillovers and productivity growth. *Applied Economics Letters*, 3: 763–767.

Bhatnagar, P. & Srivastava, R. (2003). Gravity-fed drip irrigation system for hilly terraces of the northwest Himalayas[J]. *Irrigation Science*, 21(4): 151–157.

Bitzenis, A. (2003). Universal model of theories determining FDI. *European Business Review*, 15(2): 94–105.

Bloch, H., Rafiq, S. & Salim, R. (2015). Economic growth with coal, oil and renewable energy consumption in China: Prospects for fuel substitution. *Economic Modelling*, 44: 104–115.

Blomstrom, M. & Kokko, A. (1996). *Home Country Effects of Foreign Direct Investment: Evidence from Sweden.* Cepr Discussion Papers.

Blonigen, B. A. & Figlio, D. N. (1998).Voting for protection: Does direct foreign investment influence legislator behavior? *American Economic Review*, 88(4): 1002–1014.

Blonigen, B. A., Davies, R. B. & Head, K. (2003). Estimating the knowledge-capital model of the multinational enterprise: Comment. *American Economic Review*, 93(3): 980.

Bloomstrom, M. & Persson, H. (1983). *Foreign Direct Investment: Strategic Transfer Through Supply Chains, Mimeo.* Berkeley: Haas School of Business, University of California.

Bloomstrom, M. & Wolf, E. (1994). *Multinational Corporations and Productivity Convergence in Mexico.* London: Oxford University Press.

Bollen, N. P. B. (1999). Real option and product life cycles. *Management Science*, 45(5): 670–684.

Brooke, J. (1998). Management and valuation of an environmentally sensitive area: Norfolk Broadland, England, Case Study. *Environmental Management*, 12(2): 193–207.

Caniels, Keilbach. (2000). Knowledge Spillovers and Economic Growth: Regional Growth Differentials Across Europe[M]//Knowledge spillovers and economic growth: Regional growth differentials across Europe. E. Elgar.

Cantwell, J. & Santangelo, G. D. (1999). The frontier of International Technology Networks: Sourcing abroad the most Highly Tacit Capabilities. *Information Economics and Policy*, 11(1): 123.

Cantwell, J., Iammarino, S. & Noonan, C. (2001). Sticky Places in Slippery Space — The Location of Innovation by MNCs in the European Regions. *Inward Investment Technological Change and Growth.* London: Palgrave Macmillan UK.

Cao, Y. B. & Guan, J. C. (2000). An empirical study on fuzzy comprehensive evaluation method for decision to suspend R&D project. *R&D Management*, 12: 21–26.

Carr, D. L., Markusen, J. R. & Maskus, K. E. (2003). Estimating the knowledge-capital model of the multinational enterprise: Reply. *American Economic Review*, 93(3): 995–1001.

Caves, D. W., Christensen, L. R. & Diewert, W. E. (1982). The economic theory of index numbers and the measurement of input, output and productivity. *Econometrics*, 50: 1393–1414.

Caves, R. E. (1971). International corporations: The industrial economics of foreign investment. *Economica*, 38: 5.

Caves, R. E. (1974). Multinational firms, competition and productivity in host-country markets. *Economica*, 41: 176–193.

Charnes, A., Cooper, W. W. & Rhodes, E. (1978). Measuring the efficiency of decision making units. *European Journal of Operational Research*, 2: 429–444.

Chen, C. T. *et al.* (2004). Using DEA to evaluate R&D performance of the computers and peripherals firms in Taiwan. *International Journal of Business*, 9(4): 261–288.

Chen, G. H. (1997). On decision to suspend R&D project. *Science Research Management*, 4: 75–80.

Chen, H. (2001). New ideas for promoting the upgrading of industrial structure in China with foreign investment. *South China Journal of Economics*, 4: 28.

Chen, S. J. (2006). *Research on Central Influence Factors of Multinational Corporations' Establishment of R&D in China*. Nanjing: Hohai University.

Chen, T. T. (2003). Research on the internal mechanism of spillover effects of China's FDI industry. *The Journal of World Economy*, 9: 23–28.

Chen, T. T. & Chen, J. (2006). Industrial growth factors and spillover effects in China's FDI industry. *Economic Research Journal*, 6: 61–73.

Chen, V. Z., Li, J. & Shapiro, D. M. (2012). International reverse spillover effects on parent firms: Evidence from emerging-market MNEs in developed markets. *European Management Journal*, 30(3): 204–218.

Chen, X. D. & Liang, T. Y. (2010). R&D efficiency of China's high-tech industry and its influencing factors — An empirical study based on panel data and SFPF model. *Studies in Science of Science*, 2010(8): 1198–1205.

Chen, X. T. & Yang, L. X. (1998). Attempts and reflections on evaluation of findings of fund projects. *Journal of Management Sciences in China*, 1: 97–100.

Chen, Y. H. & Qiu, Y. H. (1999). Entropy evaluation of reliability of decisions on R&D investment. *Science Research Management*, 20: 71–75.

Chen, Y. M. (2006). Re-test of technology spillover mechanism of foreign direct investment in China's manufacturing industry. *World Economic Papers*, 3: 28–42.

Cheng, B. W. (2005). Quantitative analysis of the impact of R&D on total factor productivity in China. *Science and Technology Management Research*, 6: 39–42.

Cheng, J. & Bolon, D. S. (1993). The management of multinational R&D: A neglected topic in international business research. *Journal of International Business Studies*, 24(1): 118.

Ciruelos, A. & Miao, W. (2005). International technology diffusion: Effects of trade and FDI. *Atlantic Economic Journal*, 33(4): 437.

Coe, D. T. & Helpman, E. (1995). International R&D spillovers. *European Economic Review*, 39: 859–887.

Cohen, W. M. & Levinthal, D. A. (1989). Innovation and learning the two factors of R&D. *Economic Journal*, 99(9): 569–596.

Coombs, R., McMeekin, A. & Pybus, R. (1998). Toward the development of benchmarking tools for R&D projects management. *R&D Management*, 28: 175–186.

Czarnitzki, D. & Licht, G. (2006). Additionality of public R&D grants in a transition economy: The case of eastern Germany. *Economics of Transition*, 14(1): 101–131.

Dalton, D. & Serapio, M. (1999). *Globalizing Industrial Research and Development*. Washington, DC: US Department of Commerce, Office of Technology Policy.

De Mello, L. R. Jr. (1997). Foreign direct investment in developing countries and growth: A selective survey. *The Journal of Development Studies*, 34(1): 1–34.

de Saxcé, G. & Bousshine, L. (1994). Limit analysis theorems for implicit standard materials: Application to the unilateral contact with dry friction and the non-associated flow rules in soils and rocks. *International Journal of Mechanical Sciences*, 40(4): 387–398.

Djankov, S. & Hoekan, B. (2000). Foreign investment and productivity growth in Czech enterprises. *World Bank Review*, 14: 49–64.

Dobson, W. (1997). Part I. Introduction, Chapter 1. East Asian Integration: Synergies Between Firm Strategies and Government Policies.

Doh, J. P., Jones, G. K. & Teegen, M. H. (2002). Knowledge transfer in multinational corporations. Foreign research and development and host country environment: An empirical examination of U.S. international R&D. *Management International Review*, 45(2): 121–154.

Driffield, N. (2001). The impact on domestic productivity of inward investment in the UK. *The Manchester School*, 69: 103–119.

Du, D. B. (1999). R&D internationalization of multinational corporations and their types of locations. *World Regional Studies*, 3: 41–45.

Du, D. B. (2001). *Research on Location Models of R&D Internationalization of Multinational Corporations*. Shanghai: Fudan University Press.

Du, N. (2005). The effects of imprecise probabilities and outcomes in evaluating investment options. *Management Science*, 51: 1791–1803.

Dunning, J. H. (1996). Developing countries versus multinational enterprises in a globalising world: The dangers of falling behind. *Forum for Development Studies*, 26(2): 261–287.

Dunning, J. H. (1977). Trade, location of economic activity and the MNE: A search for an eclectic approach. In: Ohlin, B., Hesselborn, P. O. & Wijkman, P. M., Eds. *The International Allocation of Economic Activity*. London: Macmillan.

Enos, J. L. (1962). Invention and innovation in the petroleum refining industry. *NBER Chapters*, 27(8): 786–790.

Ernst, G. G. J. & Palmer, M. R. (1998). Generation of hydrothermal megaplumes by cooling of pillow basalts at mid-ocean ridges. *Nature*, 393(6686): 643–647.

Feng, G. F. *et al.* (2006). Empirical analysis of R&D efficiency of secondary sector in China and its influencing factors. *China Industrial Economics*, 11: 46–51.

Feng, Z. X. *et al.* (2011). Government investment, the degree of marketization and technological innovation efficiency of China's industrial enterprises. *The Journal of Quantitative & Technical Economics*, 4: 3–17.

Findlay, R. (1978). Relative backwardness direct foreign investment, technology diffusion and trade. *Journal of International Economics*, 92: 1–16.

Fors, G. & Zejan, M. C. (1996). Overseas R&D Abroad: The Role of Adaptation and Knowledge-Seeking. Working Paper, Stockholm School of Economics, The Economics Research Institute.

Franko, L. G. (1978). Foreign direct investment in less developed countries: Impact on home countries. *Journal of International Business Studies*, 9: 55.

Fredrik, S. (1999). Technology gap, competition and spillovers from direct foreign investment: Evidence from establishment data. *Journal of Development Studies*, 36: 53–73.

Freeman, C. *et al.* (1973). A Study of Success and Failure in Industrial Innovation, Report on Project SAPPHO by the Science Policy Research Unit. Center for the Study of Industrial Innovation, London: University of Sussex.

Fu, H. C. (1999). Recognition of Handwritten Similar Chinese Characters by Self-growing Probabilistic Decision-based Neural Networks. IEEE World Congress on IEEE International Joint Conference on Neural Networks. IEEE.

Fu-Long, W. U. (2004). Further analysis of Chinese investment funds' herding behavior. *Chinese Journal of Management Science*, 115–117.

Funke, M. & Niebuhr, A. (2000). Regional geographic research and development spillovers and economic growth: Evidence from West Germany. *Regional Studies*, 39(1): 143–153.

Gao, L. Y. & Wang, Y. Z. (2008). R&D spillover channels, heterogeneous response and productivity: An empirical study based on panel data from 178 countries. *The Journal of World Economy*, 2: 65–73.

Ge, S. Q. (2006). New trends and effects of international direct investment. *The Journal of World Economy*, 3: 71–74.

Girma, S. & Wakelin, D. (2001). Who benefits from foreign direct investment in the UK? *Scottish Journal of Political Economy*, 48: 119–133.

Globerman, S. (1979). Foreign direct investment and "spillover" efficiency benefits in Canadian manufacturing industries. *Canadian Journal of Economics*, 12: 42–56.

Goto, A. & Suzuki, K. (1989). R&D capital, rate of return on R&D investment and spillover of R&D in Japanese manufacturing industries. *Review of Economics and Statistics*, 71: 555–564.

Granstrand, O. (2004). The economics and management of technology trade. *International Journal of Technology Management*, 27(2–3): 209–240.

Griliches, Z. (1979). Issues in assessing the contribution of research and development to productivity growth. *Journal of Economics*, 10: 92–116.

Griliches, Z. (1988). Productivity puzzles and R&D: Another explanation. *Journal of Economic Perspectives*, 2: 9–21.

Griliches, Z. (1986). Productivity, R&D and basic research at the firm level in the 1970s. *American Economic Review*, 76: 141–154.

Griliches, Z. (1980). R&D and the productivity slowdown. *American Economic Review*, 70: 343–348.

Griliches, Z. (1992). The search for R&D spillovers. *Scandinavian Journal of Economics*, 94 (Supplement): 29–47.

Grossman, G. & Helpman, E. (1995). *Innovation and Growth in the Global Economy*. Cambridge: MIT Press.

Gu, E. G., Huang, Y. (2006). Global Bifurcations of Domains of Feasible Trajectories: An Analysis of a Discrete Predator–Prey ModeL[J]. *International Journal of Bifurcation & Chaos*, 16(09): 2601–2613.

Gui, P., Xiong, W. & He, S. (2002). Relative efficacy assessment of industrial utilization of R&D resources. *Science Research Management*, 11: 54–60.

Guilloches, Z. (1964). Research expenditures, education, and the aggregate production function. *American Economic Review*, 54: 961–974.

Hakanson, L. (1995). Learning through acquisitions: Management and integration of foreign R&D laboratories. *International Studies of Management and Organization*, 25(1): 121–157.

Hakansson, H. (1993). Getting innovations out of the supplier networks. *Journal of Business-to-Business Marketing*, 1(3): 3–34.

Hall, B. & Mairesse, J. (1995). Exploring the relationship between R&D and productivity in French manufacturing firms. *Journal of Econometrics*, 65: 263–293.

Harhoff, D. (1998). R&D and productivity in German manufacturing firms. *Economics of Innovation and New Technology*, 6: 28–49.

Hayden, E. W. (1976). *Technology Transfer to East Europe, U.S. Corporate Experience*. New York: Praeger.

He, B. S., Gu, J. R. & Yan, Y. L. (1996). *Technology Transfer and Technological Progress in China*. Beijing: Economy & Management Publishing House.

He, J. (2000). Further accurate quantification of spillover effects of foreign direct investment on China's industrial sector. *The Journal of World Economy*, 12: 29–36.

Helleiner, G. G. (1975). The role of multinational corporation in less developed countries' trade in technology. *World Development*, 3: 161–189.

Hewitt, G. (1980). Research and development performed abroad by U.S.: Manufacturing multinationals. *Kyklos*, 33: 308–327.

Hines, J. R. Jr. (1996). Altered States: Taxes and the location of foreign direct investment in America. *American Economic Review*, 86(5): 1076–1094.

Hirschey, R. C. (1981). Research and transfer of technology by multinational enterprises. *Oxford Bulletin of Economics and Statistics*, 43(2): 115–130.

Hirschey, R. C. & Caves, R. E. (1981). Research and transfer of technology by multinational enterprises. *Oxford Bulletin of Economics and Statistics*, 43(2): 115–130.

Hofbauer, V. (1998). *Evolutionary Games and Population Dynamics*. Cambridge: Cambridge University Press.

Hofmann, P. (2013). International Technology Transfer within Multinational Enterprises: What the Distance to the Technology Frontier Matters. *The Impact of International Trade and FDI on Economic Growth and Technological Change*. Berlin: Springer.

Hofstede, G. (1980). Culture and organizations. *International Studies of Management and Organization*, 10(4): 15–41.

Holsapple, C. W., Lai, H. & Whinston, A. B. (1998). A formal basis for negotiation support system research. *Group Decision and Negotiation*, 7(3): 203–227.

Hong, S. (1998). An empirical study on the impact of three types of enterprises on China's industrial structure effect. *Economic Research Journal*, 1: 32–43.

Hou, H. P. *et al.* (2001). Model analysis of R&D knowledge spillover effect. *System Engineering Theory and Practice*, 9: 30–31.

Hou, X. B. (2009). Real estate price and heterogeneous investment behavior in China. *Economic Modelling*, 60: 271–280.

Hu, A. (2001). Ownership, government R&D, private R&D, and productivity in Chinese industry. *Journal of Comparative Economics*, 29: 136–157.

Hu, A., Jefferson, G. & Qian, J. (2005). R&D and technology transfer: Firm level evidence from Chinese industry. *Review of Economics and Statistics*, 87: 780–786.

Huang, J. B. & Fu, F. (2004). Empirical analysis of the relationship between FDI and Guangdong's technological progress. *Management World*, 9: 81–86.

Huang, L. C. & Luo, Y. F. (2006). *R&D Internationalization Research*. Beijing: Tsinghua University Press.

Huang, X. F. & Liu, D. (2000). Analysis of influencing factors of multinational corporations' R&D in China. *Science Research Management*, 9: 78.

Huang, X. H. (2005). Analysis of technology spillover effect of foreign trade and foreign investment in China. *Journal of International Trade*, 1: 27–32.

Hughs, T. P. (1994). *Technological Momentum: Does Technology Drive History?* London: MIT Press.

Hymer, S. H. (1960). *The International Operations of National Firms: A Study of Foreign Direct Investment*. Cambridge: MIT Press.

Jackson, B. (1983). Decision methods for evaluating R&D projects. *Research Management*, 27(2): 21–25.

Jaffe, A. B. (1986). Technological opportunity and spillovers of R & D: Evidence from firms' patents, profits, and market value. *The American Economic Review*, 76(5): 984–1001.

Jalilian, H. (1996). Foreign investment location in less developed countries: A theoretical framework. *Journal of Economic Studies*, 23(4): 18.

Javorcik, B. S. & Smarzy, B. (2004). Does foreign direct investment increase the productivity of domestic firms? In search of spillovers through backward linkages. *American Economic Review*, 94(3): 605.

Jefferson, G., Bai, H., Guan, X. & Yu, X. (2006). R&D performance in Chinese industry. *Economics of Innovation and New Technology*, 15: 345–366.

Jiang, D. C. & Xia, L. K. (2005). Empirical analysis of role of foreign direct investment in technology innovation in China's high-tech industry. *The Journal of World Economy*, 2005(8): 3–10.

Jiang, D. C. (2001). The game of technological innovation between multinational corporations and enterprises in the developing host countries. *The Journal of World Economy*, 9: 31–40.

Jiang, X. J. (2004). Understanding globalization of science and technology — Resource reorganization, advantage integration and enhancement of indigenous innovation. *Management World*, 6: 4–13.

Jude, C. (2016). Technology spillovers from FDI: Evidence on the intensity of different spillover channels. *The World Economy*, 39(12): 1947–1973.

Kaplan, R. S. & Norton, D. P. (1998). *Using the Balanced Scorecard as a Strategic Management System*. Beijing: Xinhua Publishing House.

Karlsson, M. (2004). Measuring R&D productivity: Complementing the picture by focusing on research activities. *Technovation*, 24: 179–186.

Kedia, B. L., Keller, R. T. & Jullan, S. D. (1992). Dimensions of national culture and the productivity of R&D units. *The Journal of High Technology Management Research*, 3(1): 1–18.

Keilbach, M. C. (2000). *Spatial Knowledge Spillovers and the Dynamics of Agglomeration and Regional Growth.* Heidelberg: Physica-Verlag.

Kellert. (1998). Local attitudes toward community-based conservation policy and programmes in Nepal: A case study in the Makalu-Barun Conservation Area. *Environmental Conservation*, 25(4): 320–333.

Keller, W. (2002). Geographic localization of international technology diffusion. *American Economic Review*, 92(1): 120–142.

Kerssers-Van Drongelen, I. C. & Cook, A. (1997). Design principles for the developing of measurement systems for research and development processes. *R&D Management*, 27(4): 346–357.

Kokko, A., Tansini, R. & Zejan, M. C. (1996). Local technological capability and productivity spillovers from FDI in the Uruguayan manufacturing sector. *The Journal of Development Studies*, 32(4): 602–611.

Komoda, F. (1986). Japanese studies on technology transfer to developing countries. *The Developing Economies*, 24(4): 405–420.

Krugman, P. R. (1981). Intraindustry specialization and the gains from trade. *Journal of Political Economy*, 89(5): 959–973.

Kuemmerle, W. (1997). Building effective R&D capabilities abroad. *Harvard Business Review*, 75(2): 61–70.

Kumar, N. (1995). Intellectual property protection, market orientation and location of overseas R&D activities by multinational enterprises. *Discussion Paper Series, The United Nations University*, 1: 673–688.

Kumar, N. (2001). WTO's emerging investment regime: Way forward for Doha ministerial meeting. *Economic and Political Weekly*, 36(33): 3151–3158.

Kumar, R. L. (1996). A note on project risk and option values of investments in information technologies. *Journal of Management Information Systems*, 13(1): 187–193.

Laamanen, M. J., Forsstrom, L. & Sivonen, K. (2002). Diversity of aphanizome-nonflos-aquae (cyanobacterium) populations along a Baltic Sea salinity gradient. *Applied and Environmental Microbiology*, 68(11): 5296–5303.

Lai, M. Y. & Bao, Q. (2005). Foreign direct investment and technology spillover: Research based on absorption capability. *Economic Research Journal*, 8: 95–105.

Lall, Q. E. H. (1999). Competing with labour: Skills and competitiveness in developing countries. International Labour Office: *Issues in Development, Discussion Paper 31.*

Lall, R. B. S. (1979). Readings on the multinational corporation in Kenya by Raphael Kaplinsky; Transnational corporations in Southeast Asia and the Pacific by Ernst Utrecht.

Lapan, H. & Bardhan, P. (1973). Localised technical progress and transfer of technology and economic development. *Journal of Economic Theory*, 6: 585–595.

Lee, Y.-S. (2005). Foreign direct investment and regional trade liberalization: A viable answer for economic development? *Journal of World Trade*, 39(4): 701–717.

Lee, H. & van der Mensbrugghe, D. (2001). A general equilibrium analysis of the interplay between foreign direct investment and trade adjustments. *Social Science Electronic Publishing*.

Li, A. F. (2004). *R&D Internationalization of the Multinational Corporations*. Beijing: People's Publishing House.

Li, H. X. (1994). Fuzzy mapping and F base number. *Applied Mathematics: A Journal of Chinese Universities: Series A*, 2: 177–186.

Li, J. (2005). The location model of multinational corporations' R&D in China. *International Economics and Trade*, 1: 73–75.

Li, J. Z. (2004). *Supply System of Industrial Common Technology*. Beijing: China Financial Publishing House, pp. 37–38.

Li, M. & Su, Q. R. (2001). Foreign direct investment and technology spillover effect. *New Heights*, 5: 43.

Li, P. (1999). Analysis of spillover effects in technology diffusion. *Nankai Journal*, 2: 28–33.

Li, P. (2006). The path and method of international technology diffusion. *The Journal of World Economy*, 9: 85–93.

Li, X. P. (2006). International trade, R&D spillovers and growth of productivity. *Economic Research Journal*, 2: 56–64.

Li, X. P. & Zhu, Z. D. (2006). International trade, R&D spillovers and growth of productivity. *Economic Research Journal*, 2: 31–43.

Li, X. Z. & Zhang, X. T. (2004). Analysis of technology spillover effect of foreign direct investment on industrial economy in the Yangtze River delta region of China. *Finance & Trade Economics*, 12: 75–80.

Li, Z. N. (2003). Research on influence of foreign capital on production efficiency of Chinese enterprises. *Management World*, 4: 35–43.

Lian, L. (2006). International comparative study on relationship between R&D investment and economic growth. Wuhan: Wuhan University of Technology.

Liang, L. X. & Liu, J. Q. (2004). Exploration of evaluation methods for R&D performance of high-tech enterprises. *Science of Science and Management of S&T*, 11: 29–32.

Liang, Z., Xue, L., Zhu, Q. & Zhu, X. Y. (2008). R&D internationalization and local knowledge exchange: An empirical analysis of multinational corporations' R&D institutions in Beijing. *The Journal World Economy*, 2: 3–16.

Lichtenberg, F. & Van Pottelsberghe, B. (1996). International R&D spillovers: A re-examination. *NBER Working Paper*, 5688.

Lin, J. C. & Chai, Z. D. (1998). Analysis of international trend of technology research and development of multinational corporations. *The Journal of World Economy*, 7: 35–39.

Lin. L. I. & Liu, Y. (2003). Risk investment policies for a firm with exponent utility. *Chinese Journal of Management Science*, 66–69.

Lint, P. (1998). R&D as an option on market introduction. *R&D Management*, 28(4): 279–287.

Liu, C. & Lu, L. W. (2006). Dynamic analysis of time lag effect of foreign direct investment under VAR model. *The Journal of Quantitative & Technical Economics*, 10: 101–110.

Liu, D. R. (2003). On the location selection of multinational corporations' R&D institutions in China. *International Economics and Trade*, 4: 75–76.

Löfsten, H. & Lindelöf, P. (2002). Science parks and the growth of new technology-Based Firms. *Research Policy*, 31(6): 859–876.

Lopes, F. (1998). Project appraisal — A framework to assess non-financial aspects of projects during the project life cycle. *International Journal of Project Management*, 16(4): 223–233.

Lorenzen, M. & Mahnke, V. (2004). Governing MNC Entry in Regional Knowledge Clusters. *Knowledge Flows, Governance and the Multinational Enterprise*. London: Palgrave Macmillan UK.

Lucas, R. (1988). On the mechanics of economic development. *Journal of Monetary Economics*, 7: 3–42.

Lukonen, G. (1987). Scientific and research evaluation: A review of methods and various contexts of their application. *R&D Management*, 17(3): 207–211.

Luo, C. Y. (2006a). FDI, domestic capital and economic growth. *World Economic Papers*, 4: 27–43.

Luo, C. Y. (2006b). The impact of FDI on China's private capital growth: Based on the interprovincial panel data analysis from 1987 to 2001. *The Journal of World Economy*, 1: 31–39.

Lynn, K. G., Frieze, W. E. & Schultz, P. J. (1984). Measurement of the positron surface-state lifetime for Al. *Physical Review Letters*, 1984, 52(13): 1137–1140.

MacDougall, G. D. A. (1996). The benefits and costs of private investment from abroad: A theoretical approach. *Economic Record*, 1996(36): 13–35.

Madden, G., Savage, S. & Bloxham, P. (2001). Asian and OECD international R&D spillovers. *Applied Economics Letters*, 8: 431–435.

Magee, S. P. (1977). Application of the dynamic limit pricing model to the price of *Technology and International Technology Transfer*, 7: 1–224.

Malmquist, S. (1953). Index numbers and indifference surfaces. *Trabajos de Estatistica*, 4: 209–242.

Mansfield, E. S., Teece, D. & Romeo, A. (1979). Overseas R&D by U.S.-based firms. *Economica*, 46: 313–331.

Mansfield, J. M. (1981). Genetics of resistance to African trypanosomes: Role of the H-2 locus in determining resistance to infection with trypanosoma rhodesiense. *Infection and Immunity*, 34(2): 513.

Mansfield, S. (1980). America's technological society: The next decade. *Interdisciplinary Science Reviews*, 5(3): 173–181.

Mariani, S. & Alcoverro, T. (1999). A multiple-choice feeding-preference experiment utilising seagrasses with a natural population of herbivorous fishes. *Marine Ecology Progress*, 189(3): 295–299.

Melitz, M. J. & Yeaple, S. R. (2004). Export versus FDI with heterogeneous firms. *American Economic Review*, 94(1): 300–316.

Meyer-Krahmer, F. & Reger, G. (1999). New perspectives on the innovation strategies of multinational enterprises: Lessons for technology policy in Europe. *Research Policy*, 751–776.

Mi, Y. S. (2006). Regional differences and influencing factors of China's efficiency in FDI allocation — based on panel data analysis. *Finance & Trade Economics*, 11: 84–89.

Miller, R. (1993). Global R&D networks and large-scale innovations: The case of the automobile industry. *Research Policy*, 5: 33–45.

Mishra, V. V. & Bhatnagar, H. (2009). Foreign direct investment in insurance sector in India. *Macquarie Journal of Business Law*, 6(3): 173–178.

Mohnen, P. (2001). *International R&D Spillovers and Economic Growth, Information Technology, Productivity, and Economic Growth: International Evidence*. London: Oxford University Press.

Montes, M. F. (1997). Direct foreign investment and technology transfer in ASEAN. *ASEAN Economic Bulletin*, 14(2): 176–189.

Moon, P. & Bates, K. (1993). Core analysis in strategic performance appraisal. *Management Accounting Research*, 4(2): 139–152.

Moon, F. (1996). Delivering the goods at TNT: The role of the performance measurement system. *Management Accounting Research*, 7: 431–457.

Mudambi, R. (1999). On the duration dependence of MNE investment. *International Business Organization*. London: Palgrave Macmillan UK.

Müller, G. & Nettekoven, M. (1999). A panel data analysis: research and development spillover. *Economic Letters*, 25: 178–192.

Mutinelli, M. & Piscitello, L. (1998). The influence of firm's size and international experience on the ownership structure of Italian FDI in manufacturing. *Small Business Economics*, 11(1): 43–56.

Nadvi, S. S., Van Dellen, J. R. & Gouws, E. (1995). Transcranial doppler ultrasonography in hydrocephalic children with tuberculous meningitis. *British Journal of Neurosurgery*, 9(4): 519–526.

Nandy, A. (2005). Foreign investment in developing countries. *ASEAN Economic Bulletin*, 22(2): 247–248.

National Science Foundation (NSF) (1976). *Earthquake Prediction and Hazard Mitigation Options for USGS and NSF Programs*. Final Report, Washington, DC: National Science Foundation, 1976.

Neven, D. & Siotis, G. (1995). Technology sourcing and FDI in the EC: An empirical evaluation. *International Journal of Industrial Organization*, 14(5): 543–560.

Newton, D. P. & Pearson, A. W. (1994). Application of option pricing theory to R&D. *R&D Management*, 24(1): 83–89.

Niosi, J. (1997). The globalization of Canada's R&D. *Management International Review*, 37(4): 387–404.

Nixon, B. (1997). Accounting treatment of R&D expenditure: Views of UK company accountants. *European Accounting Review*, 6(2): 265–277.

Nonaka, I., Byosiere, P., Borucki, C. C. *et al.* (1995). Organizational knowledge creation theory: A first comprehensive test. *International Business Review*, 3(4): 337–351.

Nonaka, T. (2000). Shear failure of a steel member due to a blast. *International Journal of Impact Engineering*, 24(3): 231–238.

Nosov, V. & Tseplyaeva, J. (2016). China: Economy in transition. *Economic Policy*, 3: 46–55.

Odagiri, H. & Yasuda, H. (1996). The determinants of overseas R&D by Japanese firms: An empirical study at the industry and company levels. *Research Policy*, 25(9): 67–106.

OECD (1985). *Frascati Manual*. Shanghai: Fudan University Press.

OECD (2009). Investment: Unlocking Africa's potential. *OECD Journal General Papers*, 2009(1): 37–62.

Ozawa, T. (1992). *Foreign Direct Investment and Economic Development in Transnational Corporations*. New York: United Nations.

Pan, W. Q. (2003). The spillover effect of foreign investment on China's industrial sector: Analysis based on panel data. *The Journal of World Economy*, 6: 3–7.

Papanastassiou, M. (1997). Firm-strategies and the research-intensity of US MNEs' overseas operations: An analysis of host-country determinants. *Global Competition and Technology*.

Patel, P. & Pavitt, K. (1993). The technological competencies of the world's largest firms: Complex and path-dependent, but not much variety. *Research Policy*, 26(2): 141–156.

Patel, P. & Pavitt, K. (1998). The wide (and Increasing) spread of technological wompetencies in the world: A challenge to conventional wisdom. In: Chandler, A.D., Hagstrom, P. & Solvell, O. (eds.) *The Dynamic Firm: The Role of Technology Strategy Organization and Regions*. Oxford University Press, 192–213.

Patel, P. & Vega, M. (1999). Patterns of internationalization and corporatetechnology: location versus home country advantages. *Research Policy*, 28: 45–55.

Pearce, A. R. (1999). Range estimating for risk management using artificial neural. *Networks Journal of Parametrics*, 19(1): 3–31.

Pearce, R. D. (1989). *The internationalization of research and development by multinational enterprise*. New York: St. Martin's Press.

Perez, T. (1977). Multinational enterprises and technological spillovers: An evolutionary model. *Evolutionary Economics*, 7: 169–192.

Porter, M. E. (1985). *Technology and Competitive Advantage*. New York: The Free Press.

Qin, J. B. (2003). Research on R&D Performance Measurement and Control of High-Tech Enterprises. Zhengzhou: Hunan University.

Qiu, B. Y., Shuai, X. & Pei, J. (2008). Research on FDI technology spillover channel and China's manufacturing productivity growth: Analysis based on panel data. *The Journal of World Economy*, 8: 20–31.

Qu, J. S. & Guan, J. C. (1996). Fuzzy dynamic comprehensive evaluation of decisions to suspend R&D projects. *Science Research Management*, 17: 25–28.

Rao, S. N. (2003). Operating performance of the firms issuing equity through rights offer. *Article in Vikalpa*, 2003, 28(4): 25–40.

Rebentisch, R. *et al.* (1997). Vertical and adiabatic electronic excitations in biphenylene: A theoretical study. *The Journal of Chemical Physics*, 107(22): 9464.

Reddy, S. B. (2000). Digestibility and asymmetric information in the choice between acquisitions and joint ventures: Where's the beef? *Strategic Management Journal*, 21: 191–193.

Reger, G. (2001). Internationalization of research and development in pharmaceuticals. *Changing Innovation in the Pharmaceutical Industry*. Berlin: Springer.

Riedel, J. (1975). The nature and determinants of export-oriented direct foreign investment in a developing country: A case study of Taiwan. *Weltwirtschaftliches Archiv*, 111(3): 505–528.

Robinson, S. (1990). Analysing agricultural trade liberalization with single country computable General Equilibrium Models. *The WTO and Agriculture*.

Romer, P. (1986). Increasing returns and long-run growth. *Journal of Political Economy*, 10: 35–49.

Ronstadt, R. (1977). Research and development abroad by US multinationals. *Economica*, 45(180): 428.

Rui, B. *et al.* (1998). Do firms in clusters innovate more? *Research Policy*, 27: 527–542.

Rui, G. (2005). Evaluation method for new investment projects of overseas oil-gas field development. *Acta Petrolei Sinica*, 26(5): 42–47.

Sabi, M. (1988). An application of the theory of foreign direct investment to mutinational banking in LCDs. *Journal of International Business Studies*, 19(3): 433–447.

Schmitt, M. E., Freeland, T. *et al.* (1992). Reproducibility of the roth power centric in determining centric relation. *Seminars in Orthodontics*, 9(2): 108.

Schrader, S. C. (1991). Complications associated with the use of Steinmann intra-medullary pins and cerclage wires for fixation of long-bone fractures. *Veterinary Clinics of North America: Small Animal Practice*, 21(4): 687–703.

Schrank, A. (2004). Ready-to-wear development? foreign investment, technology transfer, and learning by watching in the apparel trade. *Social Forces*, 83(1): 123–156.

Schumpeter, J. A. (1934). The theory of economics development. *Journal of Political Economy*, 1(2): 170–172.

Serape, M. G. (1999). Globalization of industrial R&D: An examination of foreign direct investments in R&D at the United States. *Research Policy*, 28(2): 303–316.

Seyoum, M., Wu, R. & Yang, L. (2015). Technology spillovers from Chinese outward direct investments: The case of Ethiopia. *China Economic Review*, 33: 35–49.

Shen, B. M. (1996). Multinational corporations and world technology transfer. *The Journal of World Economy*, 3: 48–51.

Shen, H. C. (1998). R&D investment decision-making based on option pricing theory. *Science Research Management*, 19: 45–49.

Shen, K. R. & Geng, Q. (2001). Foreign Direct Investment, Technology Spillovers and Endogenous Economic Growth. *Social Sciences in China*, 5: 82–93.

Sheng, X. J. (2005). Economic analysis of investment in air separation equipment projects. *Cryogenic Technology*, 230(4): 5–7.

Shi, X. S. *et al.* (2009). Study of regional innovation efficiency and spatial difference in China. *The Journal of Quantitative & Technical Economics*, 3: 45–55.

Shung, E. S., Stadler, D. A. & Affleck-Graves, J. F. (2005). The performance of family controlled companies on the JSE: A financial and investment evaluation. *Investment Analysts Journal*, 29: 7–15.

Simon, W. S. (1973). Energy in the USA after the President's message: Changes in investment and balance of payments. *Energy Policy*, 1(3): 187–194.

Sounder, M. & Mandakovic, T. (1986). R&D project selection model. *Research Management*, 29(4): 36–42.

Srivastava, S. (2003). Globalization and the quality of foreign direct investment. *ASEAN Economic Bulletin*, 20(2): 193.

Su, F. L. (2006). Research on spatial pattern of R&D spillover in China's provinces. *Studies in Science of Science*, 10: 696–701.

Swan, E. S. L. (1973). Monopoly and competition in the market for durable goods. *The Review of Economic Studies*, 40(3): 333–351.

Tang, D. X. *et al.* (2008). Regional difference of the impact of R&D on technological efficiency and its path dependence. *Science Research Management*, 2: 115–121.

Teece, D. J. (1986). Profiting from technological innovation: Implications from integration, collaboration, licensing and public policy. *Research Policy*, 15(6): 28–305.

Tsang, E. W. K., Yip, P. S. L. & Toh, M. H. (2008). The impact of R&D on value added for domestic and foreign firms in a newly industrialized economy. *International Business Review*, 17: 423–441.

UNCTAD (2000). *World Investment Report 1999 (First Edition)*. Beijing: China Financial & Economic Publishing House.

Vernon, R. (1966). International investment and international trade in the product cycle. *Quarterly Journal of Economics*, 153: 190–207.

Veugelers, R. & Kesteloot, K. (1995). Stable R&D cooperation with spillovers. *Journal of Economics & Management Strategy*, 4(4): 651–672.

Vincenzo, A. & Beniamino, Q. (2001). Do R&D expenditures really matter for TPF? *Applied Economics*, 33: 1385–1389.

Wang, F. (2003). Does foreign direct investment promote technological progress in domestic industrial enterprises? *World Economy Study*, 4: 56–63.

Wang, H. L. & Li, D. K. (2006). FDI and independent research and development: Empirical research based on industrial data. *The Journal of Economic Research*, 2: 44–56.

Wang, J.-Y. & Blomstrom, M. (1992). Foreign investment and technology transfer: A simple model. *European Economic Review*, 36: 137–155.

Wang, P. H. (1999). Review of evaluation methods for scientific research projects. *Science Research Management*, 3: 18–24.

Wang, S. R. (1994). Economic analysis of research and development of suspension project. *Studies in Science of Science*, 2: 45–51.

Wang, X. H. (1998). On the trend of nationalization of research and development, causes of formation and China's countermeasures. *Social Sciences in Guangdong*, 6: 61–66.

Wang, Y. W. (2006). *Quantitative Analysis of the Impact of R&D on Total Factor Productivity in China*. Wuhan: Huazhong University of Science and Technology.

Wang, Y.-Q., Zhang, Z.-M. & Zhou, W. (2004). Arithmetic research of investment multiplicator in highway construction investment. *China Journal of Highway & Transport*, 17(2): 105–108.

Wang, Z. (2003). Estimation of R&D spillover between china and the United States. *Studies in Science of Science*, 4: 396–399.

Wang, Z. P. & Li, Z. N. (2004). Foreign direct investment, spillover effects and endogenous economic growth. *World Economic Papers*, 3: 23–33.

Werner, G. & Nettekoven, M. M. (1999). A panel data analysis: Research and development spillover. *Economics Letters*, 64(1): 37–41.

Wu, C. (2005). Overseas Listing of Chinese Companies: An Overview. *China & World Economy*, (04): 46–59.

Wu, H. L. & Wu, S. Y. (2000). *Multinational Corporations' Technology Transfer to China*. Beijing: Economy & Management Publishing House.

Wu, L. H. & Wu, S. Y. (2002). *Theory of Technology Transfer from Multinational Corporations to China*. Beijing: Economy & Management Publishing House, 13.

Wu, Y. B. (2006a). R&D and productivity: An empirical study based on China's manufacturing. *Economic Research Journal*, 11: 60–71.

Wu, Y. B. (2006b). *R&D Innovation and Productivity — Evidence from China's Secondary Sector*. Beijing: Chinese Academy of Social Sciences.

Xia, J., Ortiz, J. & Wang, H. (2016). Reverse technology spillover effects of outward FDI to P. R. China: A threshold regression analysis. *Applied Economics Quarterly*, 62(1): 51–67.

Xia, X. H. & Yang, Z. S. (1997). The industrial promotion of multinational corporations and the behavior of host governments — A case study of multinational corporations' investment in Pudong. *Management World*, 2: 105–111.

Xian, G. M. & Ge, S. Q. (2000). R&D internationalization strategy of multinational corporations. *The Journal of World Economy*, 10: 3–11.

Xian, G. M. & Yan, B. (2005). The spillover effect of FDI on China's innovation capacity. *The Journal of World Economy*, 10: 18–25.

Xiao, W. & Lin, G. B. (2011). Knowledge spillover of overseas R&D capital on China's technological progress. *The Journal of World Economy*, 1: 37–51.

Xie, F. J. & Zheng, S. L. (2001). Research on the system of FDI technology transfer. *Scientific Management Research*, 1: 7.

Xie, J.-W. & Zhou, L.-J. (2009). Planetesimal accretion in binary systems: Role of the companion's orbital inclination. *Astrophysical Journal*, 698(2): 2066–2074.

Xie, W. *et al.* (2008). Analysis of R&D efficiency of China's high-tech industry and its influencing factors. *Science of Science and Management of S.&T.*, 3: 144–149.

Xu, H. (2002). Analysis of globalization of technological innovation of the multinational corporations. *Scientific Management Research*, 6: 18.

Xu, Q. R. & Zhao, J. (1991). Expert system for evaluation of scientific research projects. *Science Research Management*, 4: 23–29.

Xu, T. (2003). Introduction of FDI and China's technological progress. *The Journal of World Economy*, 10: 22–27.

Xu, X. W. & Cai, H. (2003). Research on evaluation and measurement system for regional R&D input performance. *Science of Science and Management of S.&T.*, 12: 5–8.

238   *Bibliography*

Yan, Z. H. (1995). Discussion on the internalization of R&D institutions in China. *Science Research Management*, 5: 37–43.

Yang, J. Q. & Zhou, Z. L. (2013). China's outward direct investment on the upgrading of domestic industries: An empirical analysis. *Economic Geography*, 33(4): 120–124.

Yang, L. X. (2001). Research on R&D assessment practice. *Research and Development Management*, 13: 18–24.

Yao, Y. (1998). The influence of non-state-owned economic components on technological efficiency of Chinese industrial enterprises. *Economy Research Journal*, 12: 568–574.

Yao, Y. & Zhang, Q. (2001a). Analysis of China' industrial technological effects. *Economy Research Journal*, 10: 13–21.

Yao, Y. & Zhang, Q. (2001b). Analysis of technological efficiency of Chinese industrial enterprises. *Economic Research Journal*, 10: 13–19.

Yi, H. C. (2005). Finance, investment and investment performance: Evidence from the REIT sector. *Real Estate Economics*, 33: 203–235.

Yi, X. Z. & Zhang, Y. B. (2006). Technological gap, protection of intellectual property rights and technological progress in late-developing countries. *The Journal of Quantitative & Technical Economics*, 10: 111–121.

Yi, Y. Y. *et al.* (2005). Corporate indigenous innovation: Study on behaviors of imitative innovation and evolution of market structure. *Journal of Industrial Engineering and Engineering Management*, 1(19): 15.

Yin, X. K. (1999). Foreign direct investment and industry structure. *Journal of Economic Studies*, 26(1): 38.

Young, A. (1995). The tyranny of numbers: Confronting the statistical realities of the East Asia growth experience. *Quarterly Journal of Economics*, 8: 641–680.

Young, S. (2000). Promoting trade and investment: Practices and issues. *BJU International*, 85(S3): 60–64.

Yu, N. (2005). Performance evaluation of China's R&D expenditures: System construction and empirical research (1995–2003). *Shanghai Journal of Economics*, 9: 3–14.

Yu, S. Y., Shi, W. & Lin, M. (2005). Research on spillover channel of foreign direct investment on the technological efficiency of domestic enterprises. *The Journal of World Economy*, 6: 44–52.

Yuan, C. & Lu, T. (2005). Foreign direct investment and management knowledge spillover effect: Evidence from Chinese private entrepreneurs. *Economy Research Journal*, 3: 37–45.

Zejan, M. C. (1990). R&D activities in affiliates of Swedish multinational enterprises. *Scandinavian Journal of Economics*, 92: 487–500.

Zeng, D. M., Wang, G. S. & Qin, J. B. (2003). Development of R&D performance measurement system for high-tech enterprises. *Operations Research and Management Science*, 2: 87–92.

Zhang, A. *et al.* (2003). A study of the R&D efficiency and productivity of Chinese firms. *Journal of Comparative Economics*, 31(3): 443–464.

Zhang, H. Y. (2005a). R&D Absorption capacity of secondary sector in China and foreign technology diffusion. *Management World*, 6: 82–88.

Zhang, H. Y. (2005b). Two-sidedness of R&D, foreign investment activities and industrial productivity growth in China. *Economy Research Journal*, 5: 107–117.

Zhang, H. Y., Wu, G. S. & Zhang, J. P. (2004). Estimation of China's inter-provincial physical capital stock from 1952 to 2000. *Economy Research Journal*, 10: 35–44.

Zhang, J. P. *et al.* Preparation of activated carbon with large specific surface area from reed black liquor. *Environmental Technology*, 2007, 28(5): 491–497.

Zhang, K. H. (2001). Does foreign direct investment promote economic growth? Evidence from East Asia and Latin America. *Contemporary Economic Policy*, 19(2): 175–185.

Zhang, Q. (2004). Nearly-optimal asset allocation in hybrid stock investment models. *Journal of Optimization Theory and Applications*, 2004, 121(2): 419–444.

Zhang, T. & Zou, H. F. (1998). Fiscal decentralization, public spending, and economic growth in China. *Journal of Public Economics*, 67: 221–240.

Zhang, X. D. & Li, X. Z. (2005). Estimation and analysis of total factor productivity in the Yangtze river delta region of China. *Management World*, 11: 59–66.

Zhang, Y. F. (2003). *Analysis of Technology Spillover Effects of China's Foreign Trade*. Hangzhou: Zhejiang University, 2003.

Zhang, Y. J., Liu, Z. Zhang, H. & Tan, T. D. (2014). The impact of economic growth, industrial structure and urbanization on carbon emission intensity in China. *Natural Hazards*, 73(2): 579–595.

Zhang, Y. S., Zeng, D. M., Qin, J. B. & Zhang, L. F. (2004). R&D performance evaluation system based on principal component analysis. *R&D Management*, (4) 1–6.

Zhang, Z. Y. *et al.* (2006). Empirical study on regional technology innovation efficiency in China based on SFA model. *China Soft Science*, 2: 125–128.

Zheng, D. Y., Wu, Q. S. & Li, Z. (2000). Real option evaluation method for corporate R&D projects. *R&D Management*, 12: 27–30.

Zheng, H., Hu, H. *et al.* (2009). The fuzzy-ai modeling for optimization of long-term metro vehicle repair. IN *FSKD 2009, Sixth International Conference on Fuzzy Systems and Knowledge Discovery*, 6 Volumes. IEEE, Tianjin, China, 14–16 August 2009.

Zheng, J. H. (2008). Can China's economic growth continue? *Economics (Quarterly)*, 3(7): 777–808.

Zheng, X. J. (2006). Empirical study on technological spillover of FDI in China. *Word Economy Studies*, 5: 51–57.

Zhou, J. & Chen, Z. Y. (2005). Analysis of influencing factors of R&D decentralization of multinational corporations. *Economy and Management*, 2: 35.

Zhu, P. F. & Li, L. (2006). Study of direct effect of two methods for technology introduction. *Economy Research Journal*, 3: 90–102.

Zhu, Y. W. & Xu, K. N. (2006). An empirical study on the R&D efficiency of China's high-tech industry. *China Industrial Economics*, 11: 38–45.

Zou, Y. J. (2002). International comparison of China's technology innovation system. *Journal of Nanjing University: Philosophy, Humanities and Social Science*, 6: 33.

Index

CPSIA information can be obtained
at www.ICGtesting.com
Printed in the USA
LVHW082112160920
666264LV00003B/5